A GUIDE TO
FOSSIL COLLECTING IN ENGLAND
AND WALES

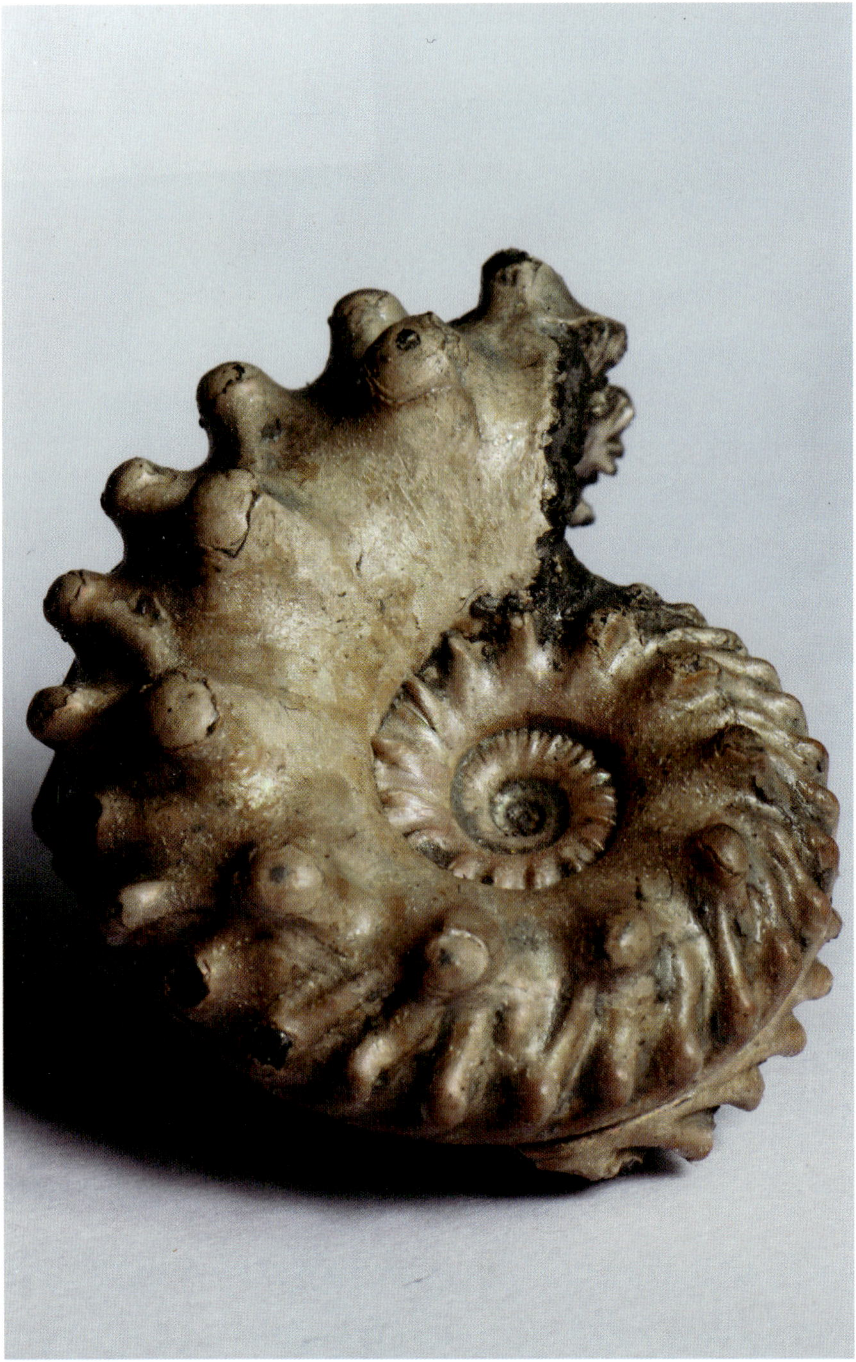

A GUIDE TO
FOSSIL COLLECTING IN ENGLAND AND WALES

Steve Snowball
&
Craig Chapman

SSP
SIRI SCIENTIFIC PRESS

Published by Siri Scientific Press, Manchester, UK
This and related titles are available directly from the publisher at:
http://www.siriscientificpress.co.uk

© 2017, Siri Scientific Press. All rights reserved. No parts of this publication may be reproduced, stored in a retrieval system or transmitted, in any form or by any means, electronic, mechanical, photocopying, recording or otherwise, without the prior written permission of the publisher. This does not cover photographs and other illustrations provided by third parties, who retain copyright of their images; reproduction permission for these images must be sought from the copyright holders.

ISBN 978-0-9929979-9-1

The authors have made every effort to contact copyright holders of material reproduced in this book and are very grateful for permissions granted. If any have been inadvertently overlooked, we apologise for our oversight which we hope will be accepted in good faith.

Steve Snowball & Craig Chapman have asserted their right under the Copyright, Design and Patents Act 1988 to be identified as the authors of this work.

Cover and book design: Steve Snowball

DISCLAIMER: This book is a general guide to fossil collecting localities and not an extensive manual for health and safety when visiting such sites. Furthermore, because potential hazards may change over time, prior to undertaking any fossil collecting activities, you need to make yourself aware of any possible RISKS, DANGERS, HAZARDS and LEGAL IMPLICATIONS associated with visiting and collecting fossils at any particular site. The publisher, authors or any associated parties cannot be held responsible for your failure to do so nor any consequences thereof. Enjoy your fossil collecting safely and responsibly.

Molar of a woolly mammoth (*Mammuthus primigenius*) from Pleistocene River Gravels, near Peterborough, found by Mark Goble.

ACKNOWLEDGEMENTS

The authors thank and acknowledge the contributions of the following people, for their help and advice in the preparation of this book:

Alister Cruickshanks, the owner and founder of UK Fossils and Director of UKGE Limited and especially for permissions for the use of various photographs.

Roy Shepherd, author and designer of the Discovering Fossils website, for his generosity in kindly providing both content and allowing the use of various photographs.

Dean Lomax, palaeontologist and patron of UKAFH, for his constructive suggestions and support and for providing the foreword to this book.

Mark Goble, Aidan Philpott, Matthew McGrane, Fabrice Saunier, Andrew Eaves, Mike Greaves, Brandon Lennon, Lee-anne Collins, Martin Curtis, Chris Strutt, Sam Caethoven, Emilia di Girolamo, Nigel Larkin, Phil Hadland, Ted Gray, Fred Clouter, Pete Lawrance, Sam Davies, Benjamin Hutchings Halmkin, Craig Chivers, Tony Gill (Charmouth Fossil Shop), Egidio Viola, Lyme Regis Museum, Oxford Museum of Natural History, Allhallows Museum (Honiton), Fairlynch Museum (Budleigh Salterton) and Sidmouth Museum, for their support in providing various photographs used within this publication.

Andy Temple for his valued contribution to the stratigraphy of the locations.

The Jurassic Coast Team, Joe Shimmin, James Turner and Paul Streather for their help and support.

Nick Hanigan for his contributions to the section on his discovery of *Dracoraptor hanigani* and Bob Nicholls at Palaeocreations for the dinosaur's reconstruction.

Regional leaders of the UKAFH team, past and present, for their enthusiasm and help at the many UKAFH fossil hunt events.

David Penney, owner of Siri Scientific Press, for his patience, help and editorial suggestions made during the process of writing this publication.

This book is dedicated to our supportive wives,
Maria Snowball and Tami Chapman and our families.

CONTENTS

Acknowledgements — 6

Foreword — 10

Introduction — 12

SECTION 1

What is a Fossil? — 17

How & Where to Look for Fossils — 20

Tools & Equipment — 23

Preparing & Preserving Your Fossils — 26

Identifying & Labelling Your Fossils — 30

Storing & Displaying Your Collection — 31

Safe & Responsible Collecting

— the National Fossil Collecting Code — 33

The Golden Age of Fossil Collecting — 36

SECTION 2

Guide to Collecting Palaeozoic Fossils — 47

SECTION 3

Guide to Collecting Mesozoic Fossils — 101

SECTION 4

Guide to Collecting Cenozoic Fossils — 213

SECTION 5

Museums & Galleries — 278

Bibliography – Books — 280

Bibliography – Useful Websites — 282

About the Authors — 286

FOREWORD
By Dean R. Lomax

Since the birth of palaeontology the role of the amateur *fossil hunter* has been of utmost importance in order for the science to develop, expand and evolve at a significant rate. Many incredible fossil finds have been made across the UK by fossil hunters of all ages (including academic palaeontologists), ranging from professional collectors to children. With the latest discoveries of new and exciting fossils from some far-away exotic location, such as the vast Sahara Desert in Africa, under the ice in Antarctica or in the Badlands of the USA, fossils and especially fossil hunting in the UK is often forgotten, unknown, or simply cast aside. Yet, some of the greatest fossil discoveries ever made have been found right here in the UK. Palaeontological discoveries in the UK provided a platform for the very first scientific studies, resulting in the descriptions of new locations, new species and a vast array of different types of fossils, ranging from three-dimensionally preserved trilobites to the first identified dinosaurs.

The geology and palaeontology of the UK are exceptional in being incredibly diverse. Rocks of almost every geological period are exposed from all three eras: Palaeozoic, Mesozoic and Cenozoic, and have been studied for hundreds of years. The study of the many millions of fossils that originated from the rocks throughout the UK has greatly improved our understanding of palaeontology, which has provided us with a unique glimpse into deep time.

Fossil hunting is for everybody. It is one of the most actively stimulating and fun aspects of palaeontology. It allows us to 'travel back in time' – to a variety of geological ages – to collect, care for, understand, examine and study fossils and to place them within the context of the palaeoenvironments in which they once lived. For a *fossil hunter*, there is no better feeling than finding a fossil and being the very first person (ever) to hold that multi-thousand or multi-million-year-old specimen.

This book fills a long-standing void. There is no other book that covers more than 50 UK fossil hunting localities in such detail, providing photographs, comments on geology, history and, most importantly for fossil hunters, access and safety information. What Steve and Craig have achieved with this wonderful book echoes their passion and genuine love for the subject, which is shared with you (the reader). The book will no-doubt appeal to amateurs, professionals and students and will likely be a starting point for any future UK palaeontologist.

To the many, many people who will read this book:

Continue to enjoy 'the hunt' – you never know what you might find or where it might lead you.

Dean R. Lomax

www.deanrlomax.co.uk

Palaeontologist & Honorary
Visiting Scientist,
The University of Manchester, UK
Patron, UKAFH

INTRODUCTION

We both share a real passion for collecting fossils and interpreting the past, so much so, that we were prompted to consider writing a book on the subject. So here is the result, a book written for amateur collectors of all ages, in the hope that some of our own passion for fossils might rub off and inspire others to start a collection of their own, or perhaps to develop the one they already have with more clarity and purpose.

Undoubtedly, a large number of books are available on the subject of fossil collecting but many are now sadly out of date, which can be unhelpful. Many of the sites described are now no longer accessible to the public. Many inland pits and quarries, that once graced our landscape, are now long closed and either filled with water, turned into 'country' leisure parks or used as landfill sites, to be swiftly followed by the construction of 200 new homes on top of the previously exposed fossil beds!

We hope to rectify this dilemma, with this new guide. The sites we have chosen are completely up-to-date regarding the geology, the stratigraphy, the exposures, the best place to look for fossils at the location and what to look for once you are there. This has been quite a challenge; the names of rock types and their formations change frequently as geologists attempt to form an internationally recognised timescale and a composite geological time chart. Our source of reference has been the British Geological Survey chart, which will be updated as improved timescales become available. Similarly, the British Geological Survey Lexicon of named rock units has been the reference source for all rocks and formations used within the stratigraphic charts in this book and were current at the time of going to press.

UKAFH (UK Amateur Fossil Hunters) which we head, has evolved into a national organisation and the largest, most popular club for amateur fossil enthusiasts in the UK. With a vast program of fossil hunts across the UK, we have used our experiences to put together this book, to help direct people to some of our most productive fossil collecting localities across England and Wales. At UKAFH, we always aim to ensure that our field visits cater for everybody, whether you are an amateur beginner or a hardened hand. We hope this book reflects this aspect too, along with our ethos of safe and responsible collecting.

Geology is important too and gives insight into the history of the Earth, providing evidence of the evolutionary history of life and past climates. If you seriously intend to collect fossils, then you will need to know something about how they are formed and about the rocks in which they can be found. The rocks will help you to interpret the fossils, to date them and provide the context of the environment in which the animal or plant once lived. That is the exciting bit! Finding a fossil bivalve in a piece of limestone rock is all very well, but to know that it was, perhaps, part of a huge colony of other molluscs, living in shallow waters of a warm Jurassic sea, which was shared with huge marine reptiles, like plesiosaurs and ichthyosaurs, is the fascinating part. If you are not fired up by that, then perhaps collecting train numbers is possibly a better bet!

In an age of rapidly advancing and evolving technologies, and where increasingly more people are working in an indoor environment, fossil collecting can be a rich, rewarding and healthy hobby. Starting up or building on your collection is just part of it. It also gets you out into the open air, savoring some of Britain's most beautiful scenery and countryside locations, with an opportunity to meet like-minded enthusiasts.

We have not included Scotland and Northern Ireland in this book. UKAFH has not conducted field events in these areas and collecting in Scotland requires permission at many sites and at some localities fossil collecting is forbidden. The large amount of unfossiliferous, igneous rock found in both Scotland and Northern Ireland also puts the countries beyond the remit of this book, which is aimed at a productive day out, with comparative ease of access and some potential for finding fossils.

Finally, you will notice that the vast majority of locations within the book are now designated as SSSIs (Sites of Special Scientific Interest). Please enjoy collecting fossils safely and responsibly and always follow the Fossil Collector's Code at all sites visited.

Steve Snowball & Craig Chapman

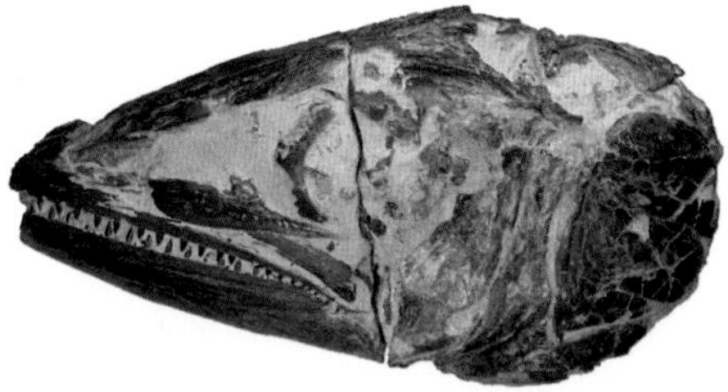

COLLECTING FOSSILS

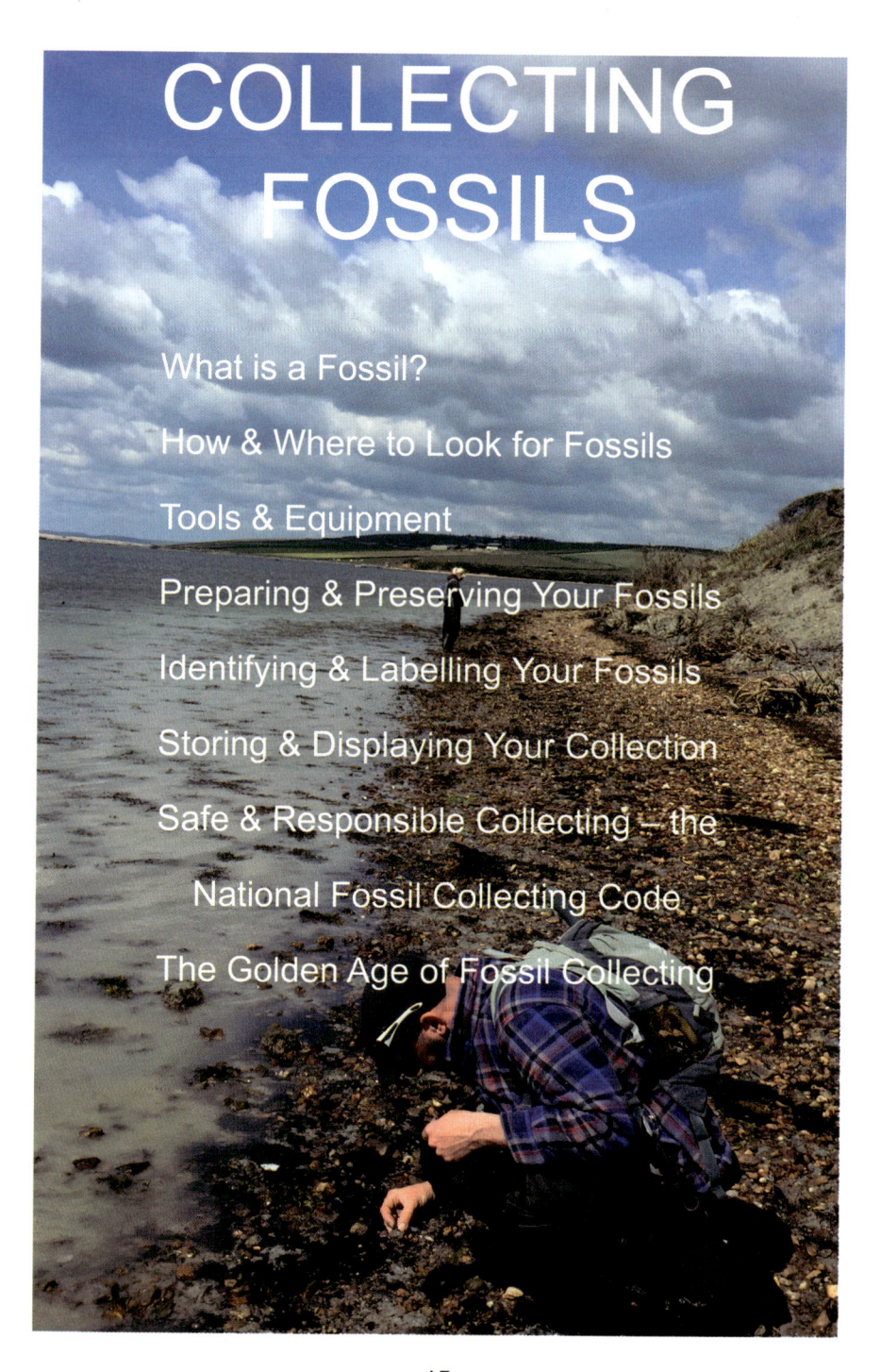

What is a Fossil?

How & Where to Look for Fossils

Tools & Equipment

Preparing & Preserving Your Fossils

Identifying & Labelling Your Fossils

Storing & Displaying Your Collection

Safe & Responsible Collecting – the National Fossil Collecting Code

The Golden Age of Fossil Collecting

An ammonite (*Euhoplites bucklandi*) from the Gault at Folkestone, Kent, found by Mark Goble.

WHAT IS A FOSSIL?

If you are considering collecting fossils, or indeed have already begun, it is worth knowing what a fossil is and how they became preserved. Fossils are the preserved remains or traces of animals, plants and other organisms of past life on Earth. The word fossil is derived from the Latin word *fossus*, which literally means 'something dug up'.

The process called fossilisation can take a very long time but fossilisation does not have to be complete to meet the definition of a fossil. Fossilisation is a continual chemical process that never stops and organisms do not have to be of a certain age in order to qualify as a fossil, although 11,700 years is now the accepted starting point.

There are several types of fossil preservation. The first type is actual preservation of the organism, where there has been a prevention of decay and bacterial action. Although relatively uncommon, such preservation is possible and good examples would be the remains of the mammoth or wooly rhinoceros found in ice and preserved without any chemical alteration. Peat bogs and oil pits can also yield such specimens.

Fossilisation by mineralisation is where all the air spaces and cavities are refilled by mineral matter. This is a common occurrence in shell and bone. Here, the mineral is brought, in solution, by the water which percolates through the earth. When precipitated out of the solution it fills up the fossil without changing the original shape or substance.

Replacement is probably the most common method. In this case, the original animal or plant substance is gradually dissolved and is replaced by a different mineral. Silica is a common replacement for wood, whereas quartz, calcium carbonate, iron pyrites and other minerals commonly replace shells and corals.

Carbonisation is a process whereby all the elements or organic materials undergo decomposition. This leaves only a residue of carbon to record the organism. This is a common process of fossilisation in plants.

Fossils found as casts, moulds or imprints are closely connected with their formation. Fossil sea urchins are often found, filled with flint, which is an example of an animal whose outer shell has been dissolved away. The cavity in the rock in which it was embedded, in this case Chalk, was filled with flint, a hard, sedimentary cryptocrystalline form of the mineral quartz, categorised as a variety of chert, leaving a sea urchin cast within the mould.

Tracks of animals are also considered to be (ichno)fossils. Footprints are often preserved in soft mud, which has hardened into rock over time. Fossil burrows, dinosaur footprints and even the impressions of jellyfish, resting on the seabed, are other examples of fossils of this type.

Fossils can also be found within the fossil excrement of other animals. The excrement is known as a coprolite, which often provides evidence of the animals eating habits, as small pieces of bone, teeth, scales and other material can be sometimes found within it.

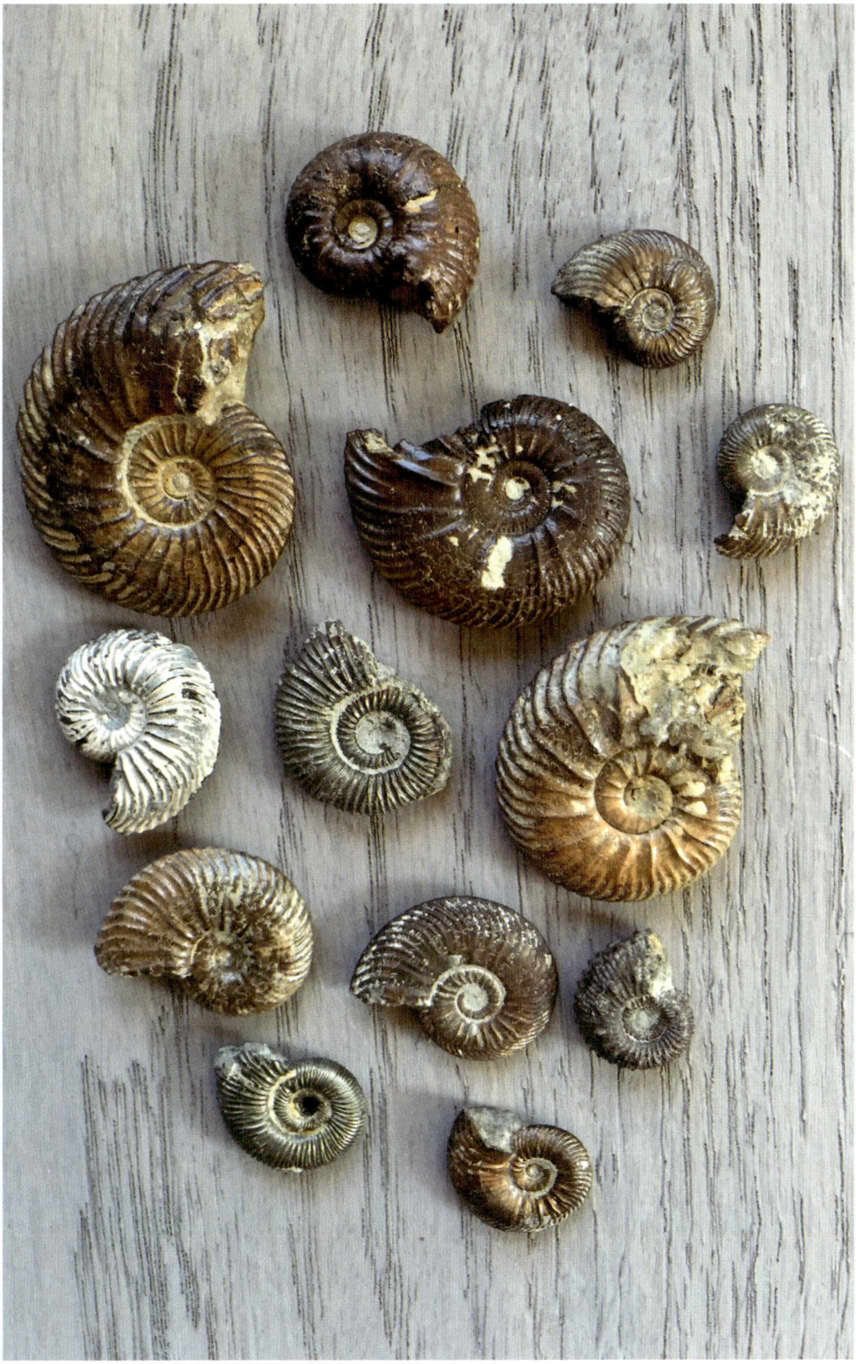

HOW & WHERE TO LOOK FOR FOSSILS

People begin collecting fossils for a number of reasons but it is certainly a rewarding hobby, increasingly enjoyed by all types of people, across all generations and all year round. Finding your own fossil is far more rewarding than buying one off of the internet and with a few basics, anyone can get involved and begin building an interesting collection of animals and plants from bygone times.

For many, collecting fossils is a means to belong to a community and certainly joining a fossil club is a good way to start, as the club is invariably educational and instructional and will most probably run regular organised field events at fossil bearing sites. For those new to collecting fossils and those already considering themselves as serious collectors, the internet can also provide forums and discussion groups, which can help with the identification of your finds and give best advice.

However, understanding just where to look for fossils is an important first step. Fossils are most commonly found in sedimentary rocks, which are a result of sediment accumulation on the bottom of a river or lake, or more often, the seabed. Clays, sandstones, limestone and shale are the most common sedimentary rocks and where fossils are most likely to occur.

Over time, these sedimentary rocks may be subject to changes taking place on the Earth's crust. As a result of intense pressure, folding, mountain building, tectonic plate movement and heat, the characteristics of the sedimentary rock may considerably alter. These are known as metamorphic rocks and a good example of one is slate, which is shale (a sedimentary rock) that has been extensively altered, along with the fossils it contained. Fossils found in metamorphic rocks are invariably distorted or flattened as a result.

Consulting a geological map will allow you to see the age and type of rocks at the surface. This is a good start, as an understanding of where sedimentary rocks or metamorphic rocks occur and their age is important. Geological maps are colour coded to help you understand which geological period the rocks represent. This in turn will help you to understand what types of fossils are likely to be found in the various rocks. For example, the shales which form the Ordovician rocks in Wales will not contain ammonites, as these animals only appeared in the Devonian and were extinct by the end of the Cretaceous.

Geological map of the Hastings area.

Once you have a basic grasp of the rock types across the UK and an overview of what fossils are potentially to be found in them, a more detailed geological map of a particular area (as shown in the photo above) and possibly the one you wish to visit, is the next step. This will show even more detail of the rock types in the region. Knowing the rock type from which your fossils are found will accurately determine their age and will help with their subsequent identification. Regional geological maps of the UK are now published by the British Geological Survey and are available online for free or as a mobile phone app.

Once you have a grasp of the basic geology of the area you wish to visit, the next step is to research those locations where the bedrock is exposed. This can often be frustrating, as the rocks are more than likely going to be obscured by topsoil at any inland outcrop and in many cases, quarries, pits and road cuttings are no longer in existence, or may be inaccessible. Coastal localities fair better, but as with any fossil collecting trip, consideration needs to be made for both personal safety and that of others. Tides, weather, mudflows, rockfalls and slippery wet rocks are important considerations at any coastal location, whether you are looking for fossils or not.

Of course, the internet provides much information about suitable locations but any accompanying issues surrounding permissions to enter quarries or private land is taken care of by an organised group. Geological societies or fossil hunting clubs will take care of public liability insurance, permissions and risk assessments too. Moreover, they will be able to give you proper advice and instruction.

We recommend UKAFH (UK Amateur Fossil Hunters) and Discovering Fossils, who are probably the two largest and best known providers of organised fossil hunts to UK locations. In addition, there are a good number of local geological societies and other fossil clubs that might also be worth considering, especially if they organise field trips and collecting visits. Organised clubs are also often able to gain access to fossiliferous sites not usually open to the general public or individuals.

TOOLS & EQUIPMENT

The list of tools and equipment recommended by many websites is often dauntingly long and can be expensive. If you purchased the complete list, you would be struggling under the weight of your rucksack before you have added all the rocks, fossils and your packed lunch! So, a good starting point is a strong, waterproof, comfortable rucksack, with various pockets, followed by a good, strong pair of walking boots with ankle support.

Of course, not every fossil collecting site requires much equipment or many tools. Many sites require little more than a pointed trowel, especially for collection in clays, gravels or soft matrix, from which fossils can be carefully retrieved. At such locations, the addition of a few collection boxes, with adequate wrapping material will suffice. There are wonderful collection boxes available, with individual compartments for your finds but for those on a tight budget, plastic containers with a lid (such as you might get with your takeaway), with tissues or bubble wrap for packing and protecting your finds, are just as good.

However, once you move onto collecting at locations where rock is hard and needs to be broken or split, you will need a geological hammer. These are essential. Old hammers found in the toolbox at the back of grandfather's shed are likely to break, splinter or cause injury to you or to others. You will have to bite the bullet and invest in a suitable hammer that is likely to give you many years of service. The price range is wide.

Many collectors are happy with a good quality brick hammer but do not waste your money on one with a tubular shaft. Ensure that it has a good solid steel shaft with a comfortable handle. Prices can start low but you pay for what you get! Geological hammers also vary in price and a quality Estwing will certainly set you back but it is considered to be the best available and will last a lifetime, provided you do not leave it on the beach!

If you are going to come across the need to break rock matrix, you will need a chisel or two. Do not use wood chisels. They are dangerous and unsuitable for geological work. They will break and cause injury. Buy a wide cold chisel, often referred to as a bolster chisel, which is ideal for splitting rock, especially shale. Also buy a finer cold chisel for extraction and prep work at home. Prices will vary but again, Estwing are the Rolls-Royce of chisels! If you are using a hammer and chisels, always wear protective eye wear, such as safety goggles or safety spectacles. Remember, one stray chipping can cause the loss of sight. We recommend that onlookers also wear the same.

Bring along a notebook and pencil, to take any notes that might be useful when labelling your find later. A digital camera and scale bar are also useful additions to your collecting kit. Also ensure you have adequate amounts of wrapping material. Bubble wrap is light, cheap and very functional. Tissues or a roll of kitchen towel can be good to take on your field trip too.

Buy a hard hat (many clubs now insist on them being worn on site) and a high vis jacket or waistcoat for quarry visits or to coastal locations and you are ready. Both are very cheap.

As your become familiar with the collection of fossils from different rock types, your basic equipment needs will probably change. Additional field equipment can be purchased for specific locations. Hand trowels are good for general digging in clays, sands, soils and crags or sifting through gravels.

Preparation of fossils back at home will also require a different set of tools for extraction, cleaning and preparation ready for display. We would recommend the prepping kit sold by UKGE and available through their website at www.ukge.com. Indeed, a large number of geological tools and equipment, books and other publications are also available through this company.

Using an air pen and compressor on an *Ichthyosaurus* skull. Image kindly supplied by Craig Chivers.

PREPARING & PRESERVING YOUR FOSSILS

One of the commonest questions we get asked when leading UKAFH fossil hunts is "How do we prepare and preserve our fossil finds?" In this section we explain some of the more simple methods.

Many fossils will require little being done to them. However, if they have been obtained from a coastal location, they will be significantly contaminated with salt and a process of desalination is necessary. Left untreated, the fossil will disintegrate, due to the salt crystals that have formed, either on the fossil or the surrounding matrix. Many fossils can to be soaked in a bowl of regularly changed tap water over a period of several weeks, but an easier method, especially for smaller specimens, is to tie them inside an old pair of tights and drop it into your toilet cistern. Here, a regular flushing and change of water will occur. After a few days you can remove the fossil and allow it to dry naturally.

Never attempt to soak fossils preserved in shale or mudstone, as they may disintegrate. Ammonites found in the Gault or shells found in the Barton Beds, will also suffer the same fate.

Of course, there are many fossils that will require no preparation at all. Shark teeth and fossils found in flint are examples of specimens that will require no treatment other than a wash and dry. Drying has to be accomplished properly. Merely placing wet fossils on a radiator or in a direct heat source is one way to disaster, as the fossil is highly likely to shrink, crack and break. Allow your fossil to dry naturally, away from direct sunlight, which may cause algal growth.

Once the process of soaking and drying is complete, use a small brush (e.g. an old tooth brush) to remove any remaining, unwanted debris and reveal more of the specimen. This can be aided by the careful use of a dental pick, to remove stubborn matrix still attached. In the case of many shells found at locations such as Barton-on-Sea, Walton-on-the-Naze or Gault ammonites from Folkestone, this dry preparation will be enough. More delicate specimens can benefit from a 1:3 coating of PVA:water, to stabilise them further.

Any cracks that appear in the fossil or matrix should be consolidated, using a fast drying liquid superglue, but take care! The idea is to get the fossil stabilised, clean, dry and ready for any further prepping that might be required and where the attached matrix is sufficiently removed, to allow the fossil to be revealed as fully as is possible.

In many cases the equipment required is minimal and a dental pick, small penknife and a variety of adapted tools can be used. This is easily achieved on softer rocks, such as Chalk but with harder matrix may require the use of a hand power engraver (which are easily affordable and effective). The Record Power is such a professional etcher and is

Brittle star (*Palaeocoma egertoni*) being cleaned with a brush and dental pick.

Using a hand-held engraving pen on an ammonite within a chalk matrix.

Pyritised fossils such as these ammonites from the Oxford Clay are often prone to 'pyrite disease'.

A tray of fossils affected by pyrite 'rot'. Image used with the kind permission of Nigel Larkin.

is excellent for fossil preparation. It has variable speed and comes with four different tips, with spares available.

When using an engraver, your preparation of the fossil probably will not be perfect (as it takes a lot of practice), so experiment on unimportant specimens first. Work on small areas and take a regular break. The vibrations from the engraver and the level of concentration required can make tiring work.

For the more advanced collector, an air pen and compressor are quick and effective for removing matrix, but the equipment is expensive and will require good practices for your personal health and safety. Using a designated work area (such as a workshop or shed) that has good air extraction, coupled with the wearing of a good quality respirator and goggles are essential.

Many fossils are preserved in iron pyrites. These commonly include many fossils found along the Jurassic Coast of Dorset and those within the Oxford Clay and the Gault. London Clay fossils from the Isle of Sheppey, including fruits and seeds, shells and bone are also pyritised. All can ultimately suffer from 'pyrite disease' or 'pyrite rot'. Pyrite (or iron persulfide FeS_2) oxidises

Another case of pyrite disease. In this example, an ammonite specimen has completely decomposed. Image used with the kind permission of Nigel Larkin.

and forms iron sulphate ($FeSO_4$); this oxidation product is several times the volume of the original mineral and the resulting crystal growth and expansion causes the specimen to fracture and totally crumble.

Keeping fossils preserved in pyrites in dry conditions, using a sealed container with a sachet of silica gel is the only way to prevent this deterioration, but once the damage begins it is irreversible. It is possible for fossils to turn into a pile of dust in a matter of weeks! If you spot pyrite disease on any of your specimens, your safest bet is to throw them out and fast!

Pyrite oxidation can not only affect the fossil itself but can also badly damage the box in which it is kept, the labels and it can also spread to other fossils in your collection! Some collectors advocate the use of sealants, such as a varnish, to help prevent the oxidation process, by preventing air directly reaching the specimen's surface. However, such coatings are mostly not impermeable to air or moisture and pyrite disease is inevitable at some point. Whilst there are two reliable ways to treat specimens (by neutralising the sulphuric acid and removing the by-products), both are technical, require some equipment and chemicals and are beyond the ability of most amateur and indeed, professional collectors.

Chemical preparation of fossils is best left to those confident in working in this area and should be attempted only when you have a complete understanding of the composition of the matrix and the specimen itself. Although treatment with weak acid is something that the amateur might consider, it will require an identification of the matrix, to allow the appropriate acid to be chosen, in order for the acid to attack the matrix and not the specimen itself. Using acids can be hazardous and the procedure should be tested on expendable pieces of matrix and bone before embarking on a full-scale acid treatment. Always wear goggles and rubber gloves. Acid can be harmful.

For the amateur, the use of vinegar (acetic acid) is as good as anything, but ensure that you neutralise the fossil with a baking soda solution and thoroughly soak and rinse with water afterwards!

Preparation of fossils in matrix using weak acid is a lengthy process, requiring careful observation of the specimen being treated and care with the chemical being used.

IDENTIFYING & LABELLING YOUR FOSSILS

The internet is a wonderful resource to help identify your fossil finds, a process made easier if you know the rock it was found in. We still believe the series published by the Natural History Museum, London to be among the best basic set of books you will ever own and use. We unreservedly recommend the following titles:

British Palaeozoic Fossils, British Mesozoic Fossils, British Cenozoic Fossils (new revised editions are available).

Once you have identified your fossil, make sure you label it. Make a label from stiff paper or thin card and include on it, as a minimum, the group (e.g. Bivalvia) the name of the fossil (e.g. *Gryphaea dilatata*), where it was found (e.g. Weymouth, Dorset), the rock formation it was found in (e.g. Oxford Clay) and the date it was found. It would also be a good idea to keep a note of your recorded specimen, in a notebook or on your computer, to enable you build a catalogue of your collection. This information will, over time, make your collection far more meaningful to you and to others.

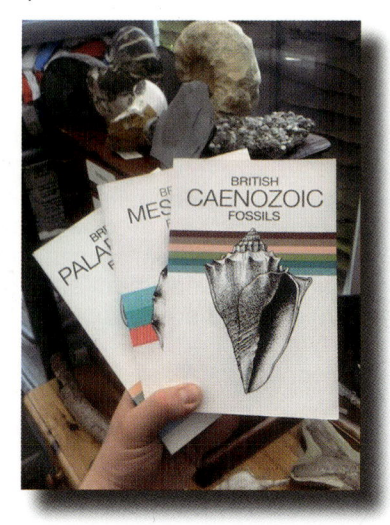

30

STORING & DISPLAYING YOUR COLLECTION

The display of your collection is very much down to personal preference and availability of space. Many people store their fossils in tray units of some sort, perhaps in a chest of drawers, or similar. The drawers can be sorted into locations or rock types (e.g. London Clay) with fossils contained within cardboard trays, which are cheap and practical. Suppliers such as UKGE sell card trays of various sizes very cheaply and these can contain both the fossil and its label.

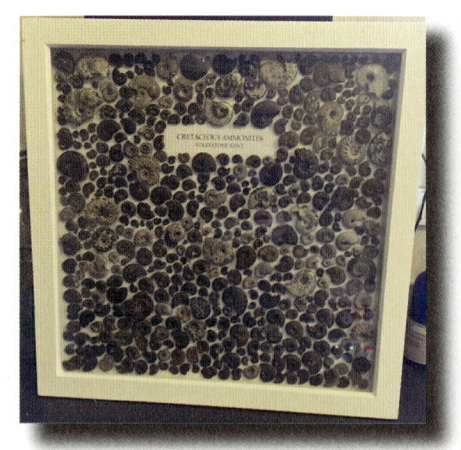

This rather artistic approach was taken by Mark Goble, who has framed his 'spare' Gault ammonites from Folkestone. His other specimens are labelled and housed elsewhere.

Some fossils, especially selected for display in a glass-fronted cabinet, rather than stored in drawers, can be a great way to show off your collection.

Bookshelf units are a common means of display, whether behind glass or not.

SAFE & RESPONSIBLE COLLECTING
The National Fossil Collecting Code

Written by Roy Shepherd of Discovering Fossils and Alister Cruickshanks of UK Fossils, the *National Fossil Collecting Code* is now widely accepted and has been taken up by fossil hunting clubs, organisations and geological societies.

PLAN
Always research the area before your visit and plan to bring the correct tools and protective equipment. You will find a vast range of information online, but you should also consider consulting a local geology group, or visiting a local library before your visit. Pay particular attention to the tools required to remove and protect specimens. In some areas fossil collecting is prohibited. Check if permission is required before visiting.

SAFETY
The nature of fossil collecting means some locations can be extremely dangerous. Before visiting a location it is highly advisable to research the potential dangers and necessary precautions. Remember to bring the correct safety equipment to protect yourself, people under your care and other people nearby. Do not take risks. Be aware of local conditions such as tides and keep away from the base of the cliff.

PATIENCE
Fossil collecting requires a great deal of patience. By researching the area before your visit, you will hopefully have the tools and equipment required to collect specimens without damaging them. Be patient and take your time. Remember, the preparation should take place at home. Whenever possible remove the specimen along with a little of the surrounding rock for protection. If you make an important discovery and do not have the correct equipment, or the find is too large. Do not risk destroying the fossil, contact your local museum for help and assistance.

RESPECT
Your initial planning should reveal the circumstances in which you may collect fossils. In many areas, collecting goes unregulated and it is therefore the sole responsibility of the collector to respect the environment. In other areas, there may be rules that govern collecting. Please accept, understand and obey any SSSI (Site of Special Scientific Interest) rules, they are there to protect the geology for future generations.

REPORT
It is important that new and significant finds are reported to the scientific community to provide an opportunity for them to be studied. We would also encourage you to report important finds to your local museum or UK Fossils, UKAFH or Discovering Fossils.

PROTECT

Fossils are often fragile, or vulnerable to damage if the necessary steps are not taken to protect them. There are two aspects to protecting your finds whilst in the field and at home. In the field, you should bring a plentiful supply of newspaper to wrap finds. You should also try to prevent them from drying out (if relevant), as soaking them at a later stage could cause fractures. A simple plastic bag will usually do the job. Once at home you should store the fossils in a safe place away from direct sunlight.

LEAVE

Not every fossil should be removed from its location. In some instances it may be too large to move, or would break in the attempt. We strongly recommend that you consider leaving the fossil for other people to see and learn from. Likewise, you should only collect a small number of specimens to allow others the opportunity of discovery.

THE GOLDEN AGE OF FOSSIL COLLECTING

In recent years, there has been a distinct upsurge in fossil collecting as a hobby. The internet has made fossils readily available for purchase and the European market is now flooded with exotic fossils from Morocco and other places. Yet the collecting of fossils is not an entirely new activity and throughout history we can trace the collection of fossils, not just for scientific purposes but also as a collecting of objects for the sake of interest, for which there was no explanation. Undoubtedly, fossil remains have fascinated man for thousands of years.

To most people, prior to the mid-nineteenth century, fossils could not be explained by orthodox wisdom. Fossils were steeped in religion, myth or legend. Country people turned to folklore to explain the inexplicable. The curiously curved *Gryphaea* oysters, for example, were seen as 'Devil's toenails'; where the Devil had trimmed his nails and left them strewn across the land! In Britain, many people were inspired by the legend of St. Hilda, an abbess who lived in the ammonite-rich area of Whitby, in North Yorkshire, during the seventh century. She rid the area of snake infestation by turning them into 'coils of stone'.

However, it was not until the late eighteenth and early nineteenth centuries that scientists began to formulate ideas about what fossils were, rather than assuming, or actually believing, that they were the work of saints or perhaps evidence of the Great Flood, as described in Genesis 6-8.

In this respect, Mary Anning (shown below, with her faithful dog, Tray), probably the most famous of fossil collectors, found herself intertwined with the fossils that she found and the scientific and cultural significance they held for the nineteenth century.

Mary's father, Richard Anning, was a carpenter, living with his family in abject poverty, in the Dorset coastal town of Lyme Regis. Food shortages in England were commonplace and the ever-rising cost of wheat forced the Anning family to seek additional sources of income. They collected and sold fossil souvenirs or 'curiosities', found on the beaches of Lyme Regis and neighbouring Charmouth, from a table set up outside their home.

Lyme Regis had started to become a fashionable seaside resort for the English aristocracy and gentry to visit, as an alternative to the unsafe travel across war-torn Europe. The visiting gentry were keen to buy the souvenir fossils and fill their cabinets full of artefacts of the natural world.

However, at this time there was still no plausible explanation for the existence of fossils, other than the Great Flood, of which fossils were seen as evidence. The dilemma that Anning faced was about the meaning of the fossils she collected and how fossils could be comprehended in light of traditional, Bible centered conceptions about the creation of life. Christian belief characterised the age. Religion pervaded social and political life and to an extent almost unimaginable today. Society tended to take refuge in the idea that religion was the only way to explain the world of nature.

A rare photograph of Broad Street, Lyme Regis taken around 1850 and much as Mary Anning would have known it. Image kindly provided by Lyme Regis Museum.

Illustration from an 1814 paper, '*Some Account of the Fossil Remains of an Animal More Nearly Allied to Fishes Than Any of the Other Classes of Animals*' by Everard Home, showing the ichthyosaur skull found by Joseph Anning in 1811.

In 1811, Mary Anning's older brother, Joseph, dug up a skull of an ichthyosaur on the ledges on Lyme Regis beach. It was a significant find and the 4-foot-long head was embellished some months later, by the rest of the skeleton found by Mary herself. The head of this strange 'crocodile', with its huge eye sockets, caused much local interest. Henry Hoste Henley, the lord of the local manor at Colway, paid the Annings £23 for the specimen.

Henley sold the skeleton on to William Bullock, a traveller, naturalist and a well-known collector of curios, who exhibited it in his London showrooms. The skull was sold again at an auction, for £45 and 5 shillings to Charles Konig, a German naturalist who was the keeper of the Department of Natural History at the British Museum, who exhibited it to the public.

Mary Anning was undoubtedly the right person, in the right place, at the right time. Geology was just developing in the early nineteenth century as a scientific discipline. Lyme Regis, with its rich fossil deposits was the right place for scientific discovery. With intense public interest, the 'Golden Age' of fossil collecting in Britain became a mixture of excitement, confusion and resistance about changing the conceptions of the history of the Earth.

William Bullock (c. 1773–1849) was an English traveller, naturalist and collector of antiquities, who exhibited the Anning's 'crocodile' skull.

Gideon Mantell was an obstetrician who lived in the town of Lewes, in East Sussex. He was an emerging palaeontologist, who had been mesmerised by the ichthyosaur specimen found by the Annings. Undoubtedly inspired by the Anning family, Mantell collected fossils and was passionately interested in studying the fossil remains of plants and animals found in his local area of Sussex.

Initially, these were specimens from the Chalk but by 1819, Mantell had begun collecting fossils from a quarry at Whiteman's Green, near Cuckfield. These fossils led to Mantell naming the rocks in which they were found the 'Strata of Tilgate Forest', which today we know to be Wealden rocks of Cretaceous age.

Gideon Algernon Mantell (1790–1852).

Mantell began to discover some large bones in the Cuckfield Quarry and in 1822, his wife Mary, also a fossil collector, found several large teeth, which Mantell could not identify. Mantell showed the teeth to other scientists, who dismissed them as those of more recent fish and certainly not of the same age as the other fossils Mantell had collected from the Tilgate Forest strata. Mantell was mocked for his finds and even Georges Cuvier, the eminent French anatomist, claimed that the teeth were merely those of a rhinoceros!

Sketch of the Whiteman's Green, Cuckfield Quarry in Sussex, with Gideon Mantell overseeing the uncovering of fossils.

Undaunted, although clearly humiliated, Mantell was convinced the teeth had come from an older Mesozoic strata of rock. He then realised the teeth resembled those of an iguana, although clearly much larger and surmised that the animal from which they had come would have probably exceeded sixty feet in length. Mantell was eventually proved correct and in 1825 he named his animal *Iguanodon*.

William Buckland, a rather quirky, enthusiastic geologist and an Anglican priest, had furiously dismissed Mantell's assertion. Moreover, he claimed the teeth were of fish. Buckland later became Dean of Westminster. He reconciled geology and an ancient Earth with religion, especially the biblical story of the flood. Geology, he believed, should be used in the service of religion.

In 1824, Buckland became President of the Geological Society and at his first meeting in this office he announced the discovery of the bones of a giant reptile, found at a quarry in Stonesfield, Oxfordshire, which he had been researching with his friend, William Conybeare.

William Buckland (1784–1856).

The animal was named *Megalosaurus*, from the Greek μέγας (megas), meaning large. Buckland had much material available to him, including a lower right jaw with a single tooth (shown below), several vertebrae, some ribs and other significantly large bones, although probably not all from the same individual. He calculated that *Megalosaurus* was about forty feet in length and was a quadruped reptile.

Buckland was not the first person to find remains of the animal in the quarry. In 1676, part of a bone had been found there and had been correctly identified as part of a femur of a large animal, which was thereafter incorrectly identified as a Roman war elephant and then changed again, as to being the scrotum of a giant human, as mentioned in the Bible!

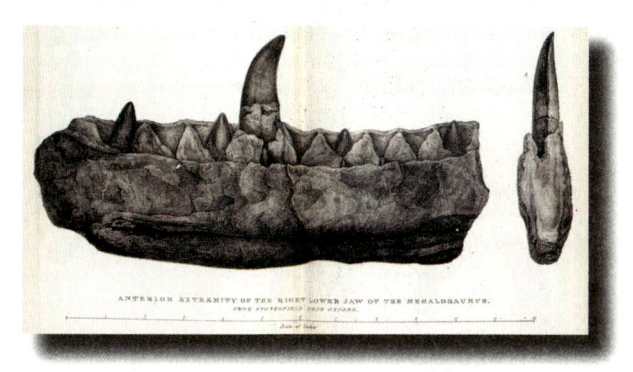

This illustration was included with William Buckland's original description of *Megalosaurus* presented to the Geological Society of London in 1824.

Back in Lyme Regis, Mary Anning continued to collect fossils and her reputation grew, especially when, in 1823, she discovered the first complete *Plesiosaurus*, followed in 1828 by the first British finding of a winged reptile, the pterosaur. Further ichthyosaurs were also discovered. Many eminent geologists and naturalists visited Anning at Lyme Regis and her visitors included Richard Owen, who attended a fossil hunt in 1839 with Coneybeare and Buckland, led by Anning.

Owen was a controversial, yet highly influential figure; a brilliant naturalist with a remarkable gift for interpreting fossils. He was not a fossil collector but a comparative anatomist. Owen had a tendency to dismiss, or ignore, the contributions made by other scientists and preferred to claim most, if not all, of the credit for himself.

Richard Owen (1804–1892). Owen was a brilliant naturalist, comparative anatomist and palaeontologist, despite being a controversial figure of the times.

Despite his initial dismissal of Mantell's *Iguanodon*, Owen was clearly inspired by the discovery and made his most famous contribution to palaeontology by his invention of the word 'dinosauria', from the Greek δεινός (deinos) meaning terrible, powerful, wondrous and σαῦρος (sauros) meaning lizard.

Frequent companions to Mary Anning were Elizabth Philpot and her sisters, who also lived in Lyme Regis. Elizabeth Philpot in particular, was well known in geological circles for her collection and knowledge of fossil fish. Her extensive collection of specimens was consulted by leading geologists and palaeontologists of the time, including William Buckland and Louis Agassiz.

The reconstruction of Iguanodon, based on a sketch by Richard Owen.

The 'Dinner in the Iguanodon Model' in 1853.

The Iguanodon creations of Benjamin Waterhouse Hawkins at Crystal Palace Gardens today.

The interpretation of fossils becoming known to science was interesting. Despite Gideon Mantell being proven to be correct in his theory about *Iguanodon* in 1825, the reconstruction of the dinosaur was far from correct. The job of overseeing the first life-size reconstruction of dinosaurs, for a newly created park at Crystal Palace, was initially offered to Mantell, who declined due to poor health. The early fossil remains were fragmentary, which led to much speculation on the posture and nature of the *Iguanodon*.

However, Owen's vision of a quadrupedal horn-nosed *Iguanodon* subsequently formed the basis on which the sculpture took shape. However, as more bones were discovered, Mantell observed that the forelimbs were much smaller than the hind limbs but his rival Owen was still of the opinion it was a stumpy creature with four pillar-like legs! The nose horn eventually was recognised as a clawed thumb.

On New Year's Eve, 31 December 1853, and immortalised in the picture published in Illustrated London News of 7 January 1854, a collection of gentlemen sat around a table inside one of the *Iguanodon* models under construction. Richard Owen is seated at the head of the table and Benjamin Waterhouse Hawkins, the sculptor, is standing centre and facing the viewer.

Buckland's *Megalosaurus* suffered much the same fate under Owen's attempts at reconstruction and the awkward reptilian quadruped is also immortalised in concrete at Crystal Palace. However, like Georges Cuvier and Gideon Mantell, Owen was making discoveries that called into question man's origins and man's importance and banishing with it the medieval concepts of Earth history.

William Buckland's *Megalosaurus*, depicted by Richard Owen as a rather clumsy quadruped.

When the Crystal Palace exhibition finally opened in June 1854, forty-thousand people gathered to see Owen's bizarre creatures, that firmly dealt a blow to the diluvian protagonists and considerably heightened public interest in natural history and of these ancient creatures, whose importance to science was now unquestioned.

In 1859, Charles Darwin published 'On the Origin of Species', probably the most important biological book ever written. It presented a body of evidence that the diversity of life on Earth arose by common descent through a branching pattern of evolution.

Darwin stressed that species formation is a gradual process and as modifications slowly develop over time, new species come to replace old ones, causing the extinction of older, less developed species.

Charles Darwin (1825–1895).

In this scheme of natural selection, the extinction of some species is inevitable. Darwin found the process of extinction mysterious, because he could not tell what types of "unfavorable conditions" led to the demise of particular species.

In 1861, Thomas Huxley was appointed to the Zoological Society Council. Richard Owen, Huxley's rival stepped down. Slowly, Owen's dinosaurs were becoming Huxley's, as his younger adversary made bold leaps forward. He began to see the links between dinosaurs and birds, after comparing *Archaeopteryx* with *Compsognathus*; something that Owen, even as a respected anatomist, had never suspected.

Huxley concluded that birds evolved from small carnivorous dinosaurs; a widely accepted theory today. Discoveries by others of fossil animals were also rapidly revealing an ancient fauna far larger and more diverse than Owen could ever have foreseen.

Thomas Henry Huxley (1825–1895).

Mary Anning died in 1847, thirty-six years after the discovery of the ichthyosaur skull; a discovery that would help change the face of science. Despite receiving no formal acknowledgement of her immense achievements during her lifetime, Anning undoubtedly helped to forge the importance of palaeontology, helped reconstruct the world's past and the history of it's life and of man's place within it.

Drawing of the skull of an ichthyosaur by Elizabeth Philpot and sent to William Buckland. It was drawn using her technique of revivifying the dried-up ink found in fossilised belemnites. The script reads:'*Drawn with colour prepared from the fossil Sepia contemporary with the Ichthyosaurus.*' Used with kind permission of Oxford University Museum of Natural History.

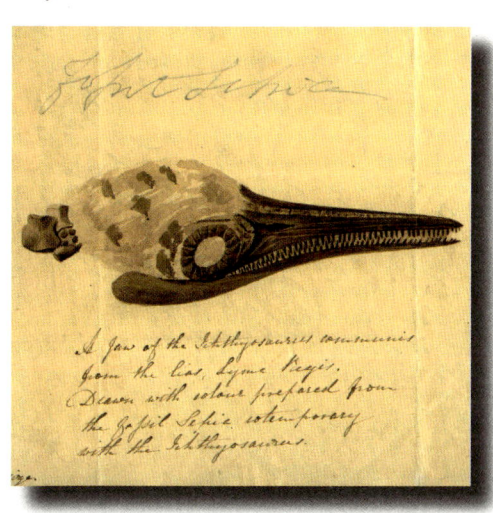

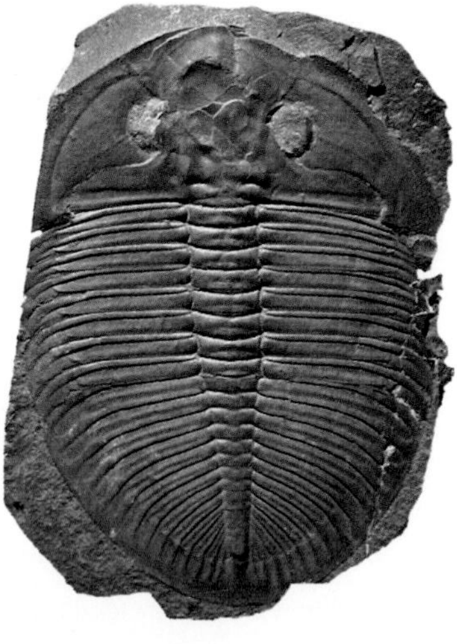

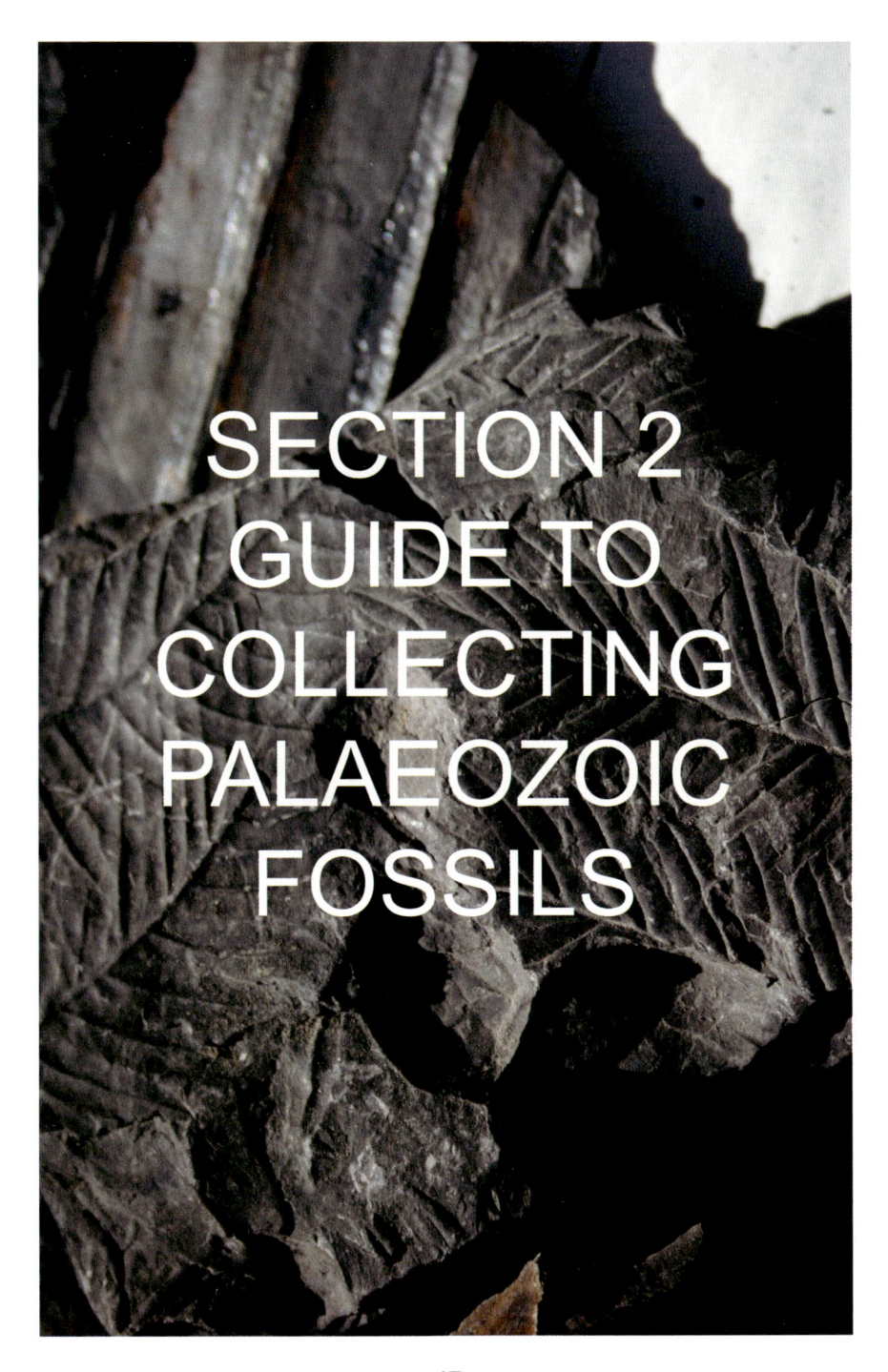

SECTION 2
GUIDE TO COLLECTING PALAEOZOIC FOSSILS

PALAEOZOIC FOSSILS

The Palaeozoic is the longest of the three geological eras. It covers 375 million years, as opposed to 155 million in the Mesozoic era and a mere 70 million in the Cenozoic. The Palaeozoic is subdivided into six geological periods: the Cambrian, Ordovician, Silurian, Devonian, Carboniferous and Permian. During the Palaeozoic, complex rocks were deposited, distorted, uplifted, tilted and folded and the preservation of the ancient fossils found in the layers are frequently distorted, yet they represent a time when life on Earth began it's diversity; from trace fossils of jellyfish, to invertebrates such as trilobites and brachiopods, to back-boned animals, land plants, insects and the forming coal seams of the Carboniferous period.

The fossil sites we have selected, fourteen in all, represent some of the most productive fossiliferous locations from the Palaeozoic era in England and Wales. In the main, Palaeozoic rocks are confined to an area west of a line running from approximately Exmouth in Devon, to near Hartlepool, County Durham and do not crop out at the surface in the southeast half of England.

Each locality has a detailed description of the site, its geology and what you can expect to find there, along with a summary chart. Please ensure that you have checked for any necessary permissions before your fossil collecting visits and at coastal locations, always check the tide times carefully and avoid collecting on an incoming tide. Some Palaeozoic locations are in rural, remote areas, so always ensure that somebody knows where you are going and when you are likely to return.

PALAEOZOIC FOSSIL SITES

The following Palaeozoic sites are described in this section, with the following colour code applied, as shown in the geological time chart below.

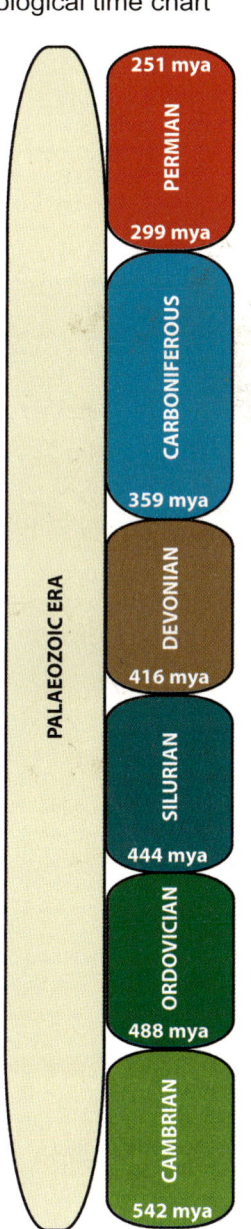

Ordovician
    Upper Gilwern Hill Quarry, Powys
    Abereiddy Bay, St. David's, Pembrokeshire
    Little Wern, Llandrindod Wells, Powys
    Gallt yr Ancr Hill, Meifod, Powys
    Druidston Haven, Pembrokeshire

Silurian
    Wren's Nest, Dudley, Shropshire
    Tites Point, Gloucestershire
    Marloes Sands, Pembrokeshire

Devonian
    Portishead, North Somerset

Carboniferous
    Druidston Haven, Pembrokeshire
    Whitehaven, Cumbria
    Betteshanger Country Park, Deal, Kent
    Newhey Quarry, Rochdale, Lancashire
    Caim, Penmon, North Wales
    Seaham, County Durham

Permian
    Seaham, County Durham

Ogyginus corndensis (Murchison), Lower Ordovician, Gilwern Hill, Builth-Wells, Powys.

UPPER GILWERN QUARRY, POWYS

On the edge of the Brecon Beacons, Upper Gilwern Hill is a site long known for its well preserved and complete trilobites. Palaeontologist, Pete Lawrance, has been studying this unique location for over 30 years and interprets the site as a near-shore, shallow water environment, used annually as a breeding and nursery colony by the trilobite *Oxyginus*.

The hill is of Lower and Middle Ordovician age and the privately owned quarry is accessible to parties and individuals upon request. The trilobite fossils to be found here are plentiful and the chances of finding a good number is very high.

The site is from the Llanvirn Series of rocks of between 467–46.9 million years of age and forms part of the Builth Inlier. The sequence of the Llanvirn Series rocks found at Gilwern is shown in the stratigraphy diagram below.

ORDOVICIAN → (DAPINGIAN) LLANVIRN SERIES →

- UPPER *Didymographus murchisoni* SHALES
- PALE FLINTY TUFFACEOUS BEDS
- MAIN RHYOLITIC TUFFS with LOWER *Didymographus murchisoni* SHALES
- RHYOLITIC TUFFS & AGGLOMERATES
- UPPER *Didym. bifidus* BEDS
- LOWER *Didym. bifidus* BEDS

A representative selection of trilobites found at Upper Gilwern Quarry in 2015 and 2016.

The fauna here is dominated by the large asaphid trilobite *Ogyginus cordensis*. Other common trilobites are the trinucleids *Bettonolithus chamberlaini* and *Trinucleus abruptus*, as well as the rarer *Anebolithus simplicior*. Other fossils include orthocones, sponges, brachiopods and graptolites.

Many fossils can be found on the scree surface but the use of a geological hammer and a bolster chisel, to split the shale along the bedding plane will reveal many and often complete specimens. Goggles are essential. The sheer number of trilobite fossils indicates a rapid burial and death, probably from hot, burning ash raining down from sub-aerial volcanoes, followed by poisoning from noxious fumes.

Access into Upper Gilwern Quarry is by way of a very isolated and rural terrain. From either Builth Wells head north, or from Llandrindod Wells head south, to Howey. Turn east, (signposted Hundred House) and after a mile there is an exceptionally steep hill (first or second gear), at the top of which is a cattle grid. You are then out onto open moorland. Take the second turning on your left (small signpost for Upper Gilwern). This is a single track road. Drive past a small cottage set down from the road and keep straight ahead. By now you will see the quarry in the distance on the hillside in front of you. Go through one gate (close behind yourself). Turn into the quarry or park next to the gate.

Upper Gilwern is an amazing site and collectors will not be disappointed at the number of trilobite specimens they can recover from this location.

Site summary

Slippery rocks
★★★★★ Accessibility
★★★★ Child suitability
★★★★ Find frequency

Upper Gilwern Quarry is on private land and has no facilities. Bring hats, sun lotion and plenty of water. This is an exposed site with no shade. Car parking is easy once the site has been found!

Suggested equipment: Geological hammer and bolster or splitting chisels, safety goggles.

Nearest postcode: LD1 5PA

Older children only. Prior permission is required and a fee payable. Contact rhiewfarm@hotmail.co.uk

Our gratitude goes to Pete Lawrance for providing much of the information about this site.

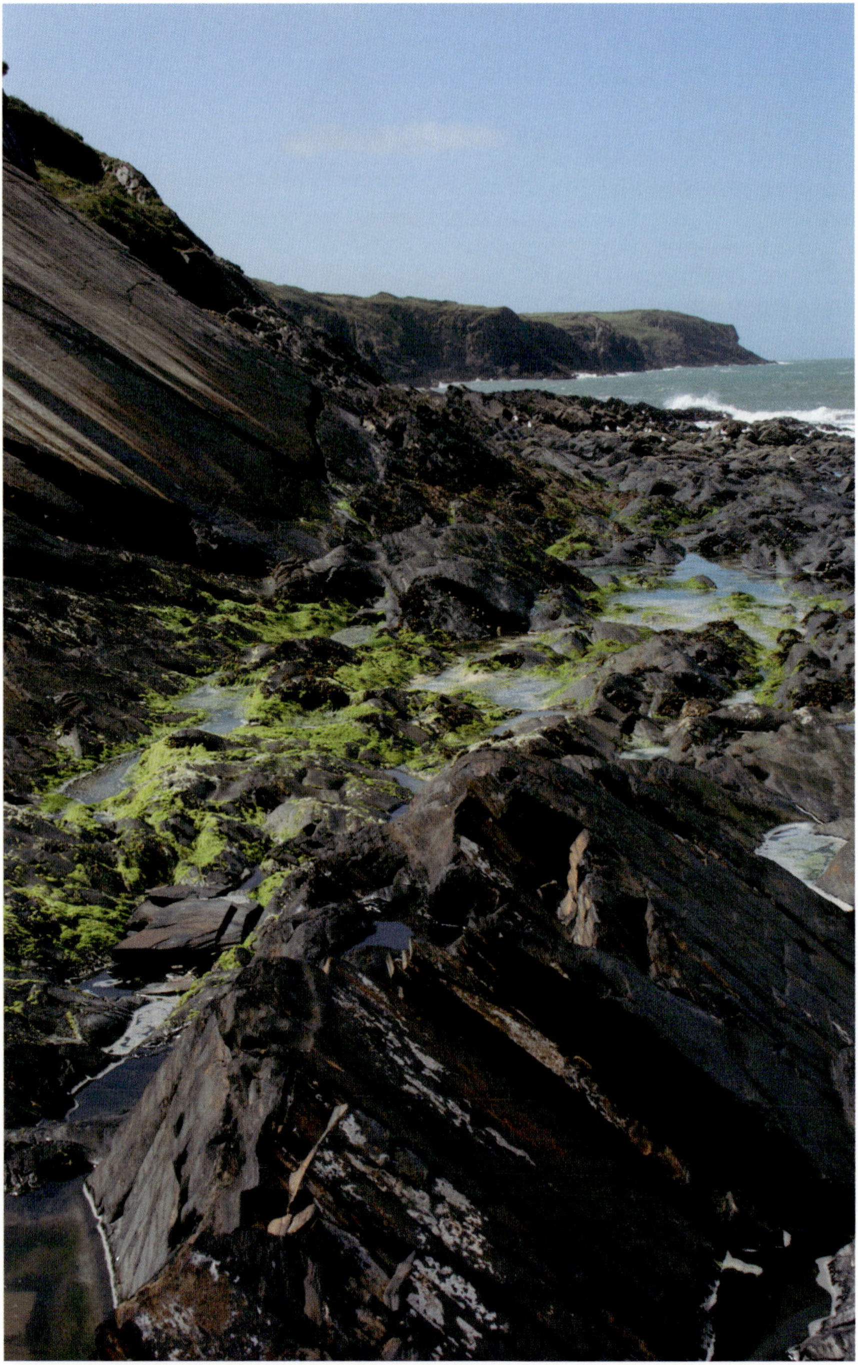

ABEREIDDY BAY, ST. DAVID'S, PEMBROKESHIRE

SSSI Site – no hammering or digging into the cliffs

Abereiddy Bay is renowned for its graptolite beds and the find frequency is high, especially among the rocks that are strewn along the foreshore. The bay is a popular tourist attraction because of the Blue Lagoon; a former quarry working that now has a beach, and a calm bay with diving points.

The fossiliferous, flaky shales are wedged within the metamorphic and igneous headland rocks but are also found on the beach. *Didymograptus murchisoni*, a graptolite shaped like a 'tuning fork' is the commonest find, along with trace fossils and the occasional planktonic trilobite (agnostids).

ORDOVICIAN — DARRIWILIAN — LLANVIRN SERIES

UPPER *Didymographus murchisoni* SHALES

LOWER *Didymographus murchisoni* SHALES

The stunning coastline along St. David's Head.

The dark shale rocks at Abereiddy Bay. Collect fossils from the foreshore and loose shale but strictly no hammering in the cliffs.

Flaky and faulted shales at Aberreidy Bay.

The shales are Middle Ordovician in age (470–464 mya) and are from the Llanvirn stage. The beach here is a mixture of pebbles and shingle. It becomes very busy during the summer with body boarders taking up much of the available space in the sea.

The sand, which is almost black and only visible at low tides, contains a high concentration of fossils and is a very popular haunt for fossil hunters. The southern end of the bay is particularly productive.

Graptolites are a fascinating group of long-extinct marine life forms. Like their contemporaries the trilobites, popular interest derives at least in part from the lack of modern descendants.

Orthograptus, a form of *Diplograptus* for example, may have been a planktonic colonial form, suspended at the water surface by a float. The area of Abereiddy Bay has long been known for its locally abundant fossils of upper Llanvirn age and local species include *Orthograptus calcaratus priscus* and *Didymograptus murchisoni*. These strata date at approximately 463 Ma and in international terms are assigned to the Darriwilian stage.

Access to Abereiddy Bay is easy, with adequate parking. Fossil collecting is possible all along the bay, from the car park along the beach. This whole area is owned by the National Trust.

Didymograptus murchisoni graptolites.

Orthograptus calcaratus priscus graptolites.

Diptograptus folium graptolites.

Site summary

⚠️ Infrequent rock falls

⚠️ Slippery rocks when wet

★★★★★ Accessibility

★★★★★ Child suitability

★★★★★ Find frequency

SSSI site.
No hammering in cliff face.

Exercise care at this site and ensure collection from the foreshore. The larger rocks can be split using a hammer and wide chisel.

Child and family friendly site. Easy access and parking, with toilet facilities. A mobile kiosk visits the car park.

Suggested equipment: Geological hammer, chisel, wrapping materials, specimen bags.

Directions: From the A487 follow the lanes from Croesgoch to Abereiddy; a slipway leads down to the beach.

Nearest postcode: SA62 6DT

LITTLE WERN, LLANDRINDOD WELLS, POWYS

This unusual quarry is set in the private grounds of 'Little Wern' holiday cottage, the rental of which allows sole access to this Ordovician trilobite site. The small quarry is from the same horizons as found at Upper Gilwern, although the trilobites tend to be much more fragile and frequently disarticulated.

Ogyginus corndensis is the most common trilobite that occurs here, although others can be found, along with other fossils, especially the graptolite, *Didymographus murchisoni*. Brachiopods and bivalves are also present at this site but are quite scarce.

ORDOVICIAN — (DARRIWILIAN) LLANVIRN SERIES

- UPPER *Didymographus murchisoni* SHALES
- PALE FLINTY TUFFACEOUS BEDS
- MAIN RHYOLITIC TUFFS with LOWER *Didymographus murchisoni* SHALES
- RHYOLITIC TUFFS & AGGLOMERATES
- UPPER *Didym. bifidus* BEDS
- LOWER *Didym. bifidus* BEDS

Little Wern quarry is an exposure of fine mudstone, of a very similar nature to that found at Upper Gilwern, laid down in the middle Ordovician about 465 mya (Llanvirnian). These mudstones show nodular features and possibly layers of bentonite, indicating local volcanic activity at the time of deposition. The deposit has also undergone compression, causing deformation and introducing cleavage at an angle to the bedding planes.

Little Wern Quarry is highly productive for trilobite fossils, the majority of which are of *Ogyginus* species, whether complete or fragmented. Unfortunately, the geologic history of this site makes the material extremely friable, so removing and preserving the majority of these specimens is quite a challenge.

It is recommended that trilobites within the thin, crumbling shale are stabilised on the spot. A 1:3 mix of PVA:water should be applied with a paint brush and allowed to dry. This can be a difficult process in the rain! The trilobite can then be extracted more safely from the surrounding matrix.

The harder, thick mudstone, that occurs here and in which fossils can also be found, is very difficult to break. Fossils within them can end up as fragments. The fragile shales dominate the left side of the quarry, whilst the rest is composed of layers of mudstone, shale and hard siderite nodules.

The privately owned quarry at 'Little Wern'.

Ogyginus corndensis **trilobite found at 'Little Wern'.**

Site summary

⚠️ Slippery rocks when wet

⭐⭐⭐⭐⭐ Accessibility

⭐⭐⭐⭐ Child suitability

⭐⭐⭐⭐⭐ Find frequency

Private quarry. Use comes with the rental of Little Wern holiday cottage.

Suggested equipment: Geological hammer, chisel, PVA and water solution (prepared in advance), wrapping materials, specimen bags.

Directions: Little Wern holiday cottage is set in a secluded spot, close to Bettwsand approximately 13 km from Llandrindod Wells. A SatNav to find the cottage is essential. Ring the owners for directions once you have reached the postcode below. To contact the owners, Mike and Maggie Ballard, Tel: 01982 570418 or e-mail: maggie.ballard123@hotmail.com

Nearest postcode: LD1 5RP

GALLT YR ANCR HILL, MEIFOD, POWYS

Many of the rugged hills around the Meifod region are composed of hard sandstone from the Ordovician of 450 mya. Gallt yr Ancr Hill, part of the Berwyns range, is typical and the highly fossiliferous beds of the Gaer Fawr Formation are exposed here, overlain by the younger sandstone rocks of the Nod Glas Formation. The Nod Glas Formation presents itself as a thin formation of black shales, with phosphatic concretions immediately below it.

The collection site is on a public footpath that passes through a wooded area and this forms part of the Glyndwr's Way. The fossils from the Gaer Fawr Formation are largely decalcified and there are death communities (known as coquinas) of many brachiopods. Crinoids can also be seen and collected from the rocks here.

ORDOVICIAN | HIRNANTIAN | CARADOC SERIES | GAER FAWR FORMATION

The public footpath into the woods allows fossil collection from stone blocks that litter the hillside and which begin to occur within a few metres of the path. These are blocks largely derived from old quarry workings. The brachiopods, *Heterorthis alternata* and *Harknessella vespertilio*, are characteristic fossils of these beds. Keeping to the path and ascending the hill slopes, fossiliferous quarry waste has also been exposed by the hooves of cattle and is worth investigating.

Sandstone blocks of quarry debris alongside the pathway at Gallt yr Ancr Hill.

The sandstone blocks have evidence of seams of brachiopods within them. These are shown by hollow brachiopod sections in the sides of the rock. When split open, these will often yield brachiopod assemblages on the bedding planes. Small trilobite pieces can also be found among the brachiopods. The right sort of blocks will yield many good quality fossils. These blocks are fairly abundant but can be somewhat difficult to find, until you get your 'eye in.'

Sandstone block from the Gaer Fawr Formation on the pathway up Gallt yr Ancr.

Practically, any sandstone blocks that are found need to be reduced in size, for carrying purposes. Also, fossils might not be apparent on the outside of the rocks, which will require splitting and investigating on site.

Use a good geological hammer and chisel and always use protective goggles when undertaking this sort of work.

Old quarry working at the summit.

Site summary

★★★★★ Accessibility
★★★★★ Child suitability
★★★★★ Find frequency

Remote location!
A mobile phone should be taken and someone should be told where you are going and when you expect to be back.

Parking can be found in a small layby next to the start of the walk into the woods at Gallt yr Ancr Hill.

Suggested equipment: Geological hammer for splitting rocks, wrapping materials, specimen bags.

Directions: Walk into the woods and after a 100 metres the hillside to the left of the gate has sandstone blocks scattered over it. Past the gate, a side path climbs into the woods to the summit of Gallt yr Ancr.

Nearest postcode: SY22 6HF
(0.57km from Gallt yr Ancr)

Various trilobite fragments and brachiopods collected from the sandstone blocks and debris along the footpath up Gallt yr Ancr Hill.

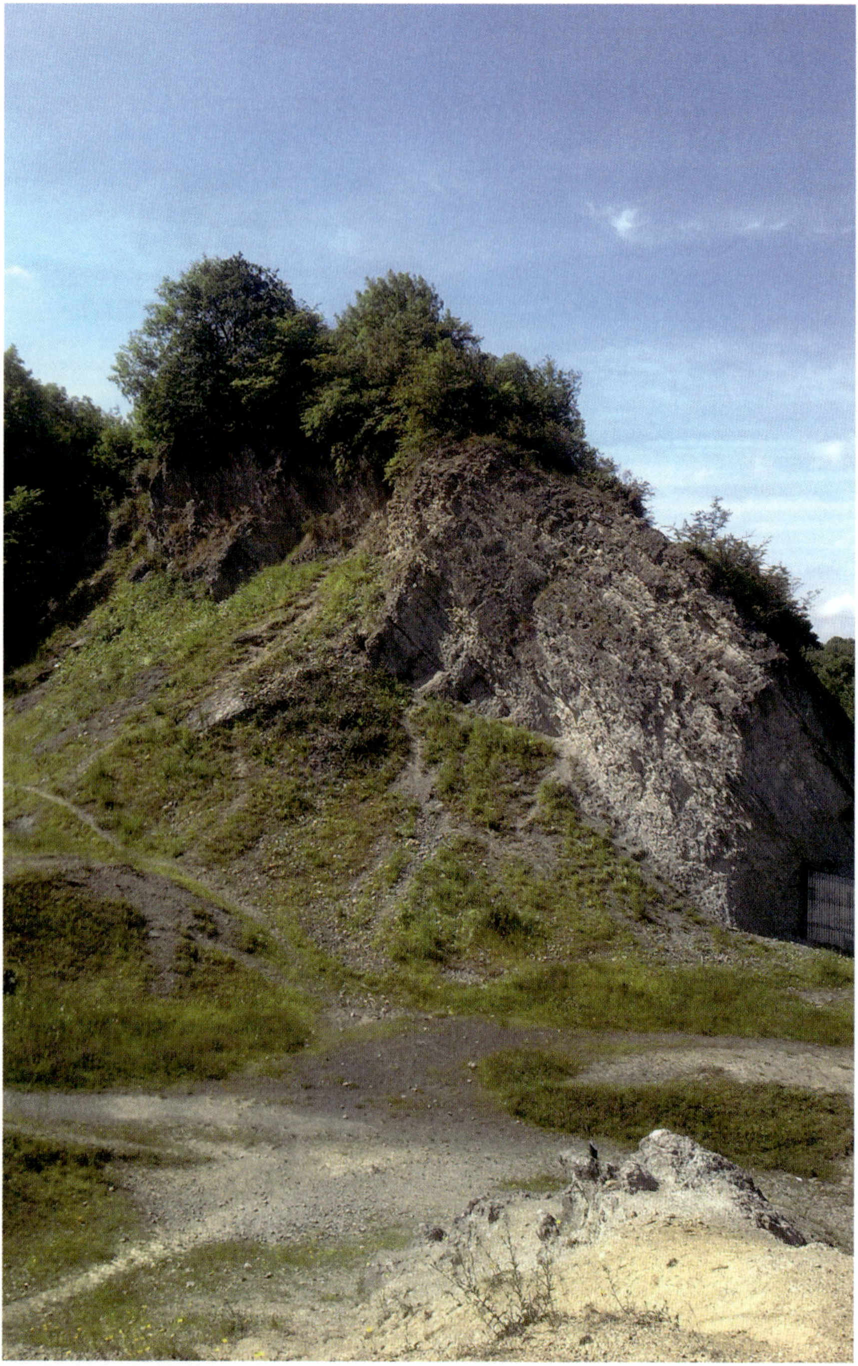

WREN'S NEST, DUDLEY, SHROPSHIRE

SSSI site – no hammers allowed on site

Wren's Nest National Nature Reserve is set to the north east of Dudley and is open to the public at all times. Wren's Nest is an Upper Silurian inlier coral reef and fossils can be collected from the loose scree at the base of the rock faces.

It is important that no hammers, chisels or other tools are used anywhere on the reserve, which has SSSI status (Site of Special Scientific Interest). In fact, no hammers are permitted on site. The beds here consist of Wenlock Limestone, dating back to the Homerian age, the early part of the Wenlock epoch of between 426–423 mya.

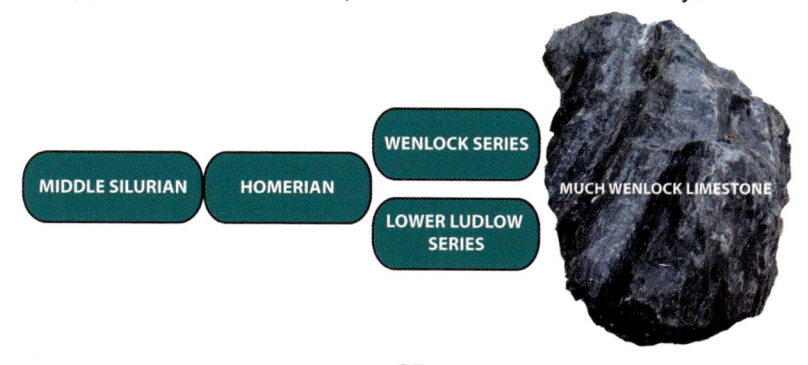

MIDDLE SILURIAN — HOMERIAN — WENLOCK SERIES / LOWER LUDLOW SERIES — MUCH WENLOCK LIMESTONE

More than 700 different types of fossil can be found here; 186 fossil species of which were first discovered and described here and 86 are found nowhere else on Earth. Wren's Nest contains the most diverse and abundant fossil fauna found in the British Isles and the fossils are among the most perfectly preserved Silurian fossils in the world.

Despite the frequency of visitors at Wren's Nest, the constantly eroding rocks and build up of scree, allows safe and productive fossil collecting. The site is ideal for children, although there are no toilet facilities within the site.

At Wren's Nest, there are so many fossils to collect that you will easily find something to take home with you. Take your time and look at the loose material on the ground. Remember do not use hammers or other tools on the rock faces. Hammers are totally banned from this site, in any case.

Fossils to be found include trilobites (especially *Calymene blumenbachii* – the 'Dudley bug', which appears on the city's coat of arms), crinoids, brachiopods (some of the easiest fossils to find on Wren's Nest are brachiopods, such as *Atrypa*, *Strophonella* and *Leptaena*), corals, sponges, bryozoans, gastropods, bivalves (especially *Goniophora* and *Pteronitella*), criconarids and nautiloids. Together they represent a superb example of a Silurian reef ecosystem.

Coral and (below) trilobite fragments, including the cephalon (head) of *Daminites*.

Site summary

⚠️ Infrequent rock falls

⚠️ Slippery rocks when wet

⭐⭐⭐⭐⭐ Accessibility

⭐⭐⭐⭐⭐ Child suitability

⭐⭐⭐⭐⭐ Find frequency

SSSI site.
No hammering or digging in cliff face.

No hammers to be taken on the site. Exercise care at this old quarry site. Rocks are stable but can still fall from above.

Contact Ian Beech, Wren's Nest Warden before collecting at this location at
ian.beech@dudley.gov.uk

Suggested equipment: Specimen bags.

Directions and parking: Wrens Hill Road, Dudley

Postcode: DY1 3SB
Car parking is available Monday to Friday 9.30am to 4pm (car park on left handside, just past the Caves Inn).

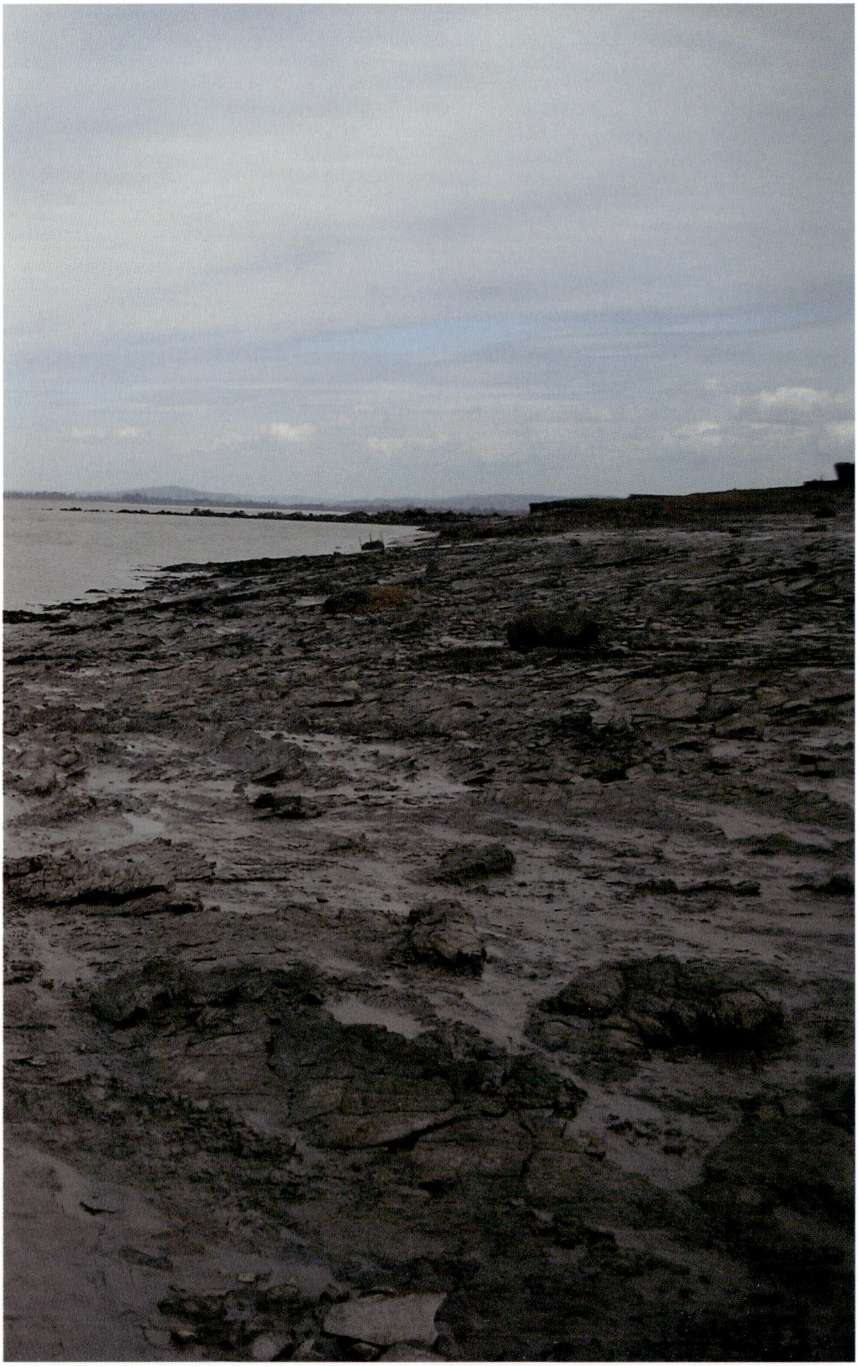

TITES POINT, PURTON, GLOUCESTERSHIRE
SSSI site – no hammering or digging into the bedrock

This site is a limited exposure of Silurian rocks, situated on the eastern shore of the River Severn, to the west of Tites Point. It is most favourable for fossil collecting during a low, scouring tide in winter months. Here, the Leintwardine (Ludlow Series) and Downton Castle Sandstone Formation of approximately 423 Ma, are exposed. The Whitcliffe Formation is a succession of silty mudstones, siltstones and thin limestone layers, the latter occurring particularly in the upper part of the sequence.

It commonly contains a concentration of brachiopods, worm tubes and fish fragments and denticles. The Whitcliffe Formation is overlain by 1.7 metres of the Downton Castle Sandstone Formation and above that, the Thornbury Sandstone. All but the tops of one or two harder bands are hidden under thick glutinous estuarine mud, which makes this site totally unsuitable for children.

- SILURIAN
- LUDLOW SERIES
- LUDFORDIAN

DOWNTON CASTLE SANDSTONE FORMATION
LUDLOW BONE BED
WHITCLIFFE FORMATION
UPPER LEINTWARDINE Fm.
LOWER LEINTWARDINE Fm.
LOWER BLAISDON BEDS

Access to this section is difficult, as the estuarine mud accumulates to a considerable depth and care needs to be exercised. In certain conditions, tides do clear the area of mud and silt, so collection is best on low scouring tides during the winter or early spring.

The site is geologically important for specimens of the heterostracan fish *Cyathaspis banksi* (an extremely rare, primitive and jawless fish, commonly found as scales), as well as *Thelodus parvidens*, *T. bicostatus*, *T. pugniformis*, *T. trilobatus* (all are characterised by a body covered with distinctive non overlapping tooth-like scales) and *Onchus* sp. (found as spine fragments). These primitive, jawless fish existed in late Silurian times, when Tites Point lay close to a shoreline and where terrestrial sediment was being fed in from a volcanic centre near the present day Mendip Hills and to the east.

Tites Point, showing the extremely muddy conditions at the site. Deep estuarine mud is prevalent to the west of the Point.

Tites Point has two distinct sections for fossil collecting. To the east of the point, the mudstones and limestone layers of the Whitecliffe Formation will reveal brachiopods, bivalves and plant remains, if split along the bedding planes. West of the Point, where the overlying Thornbury Beds meet the Downton Castle Sandstone Formation, the Ludlow Bone Beds are exposed, where fish remains are prevalent.

At this site, we cannot emphasise enough the need for care. The Severn Bore is a large surge wave, accompanied by a rapid rise in water level, which continues for about an hour and a half after the bore has passed. Ensure that you consult a tide timetable before visiting and visit on a retreating tide.

Fish vertebra.

Silurian brachiopod from the Whitcliffe Formation.

Silurian bivalve from the Whitcliffe Formation.

Site summary

Deep mud

Slippery rocks

★★★ Accessibility

★ Child suitability

★★★★ Find frequency

SSSI site. No hammering or digging in the cliffs or bedrock.

Best accessed on winter or spring, scouring low tides. Beware of deep mud (particularly western side of Tites Point) and slippery underfoot. Strong tidal flows. Double tides and beware of the Severn Bore. Not recommended for children.

Suggested equipment: Chisel pick or geological hammer and bolster or splitting chisels, safety goggles, wrapping material, specimen bag. Wellington boots are essential.

Directions: Park near to the Berkeley Arms pub in Purton village (not in the pub car park) and walk to the site by way of a public footpath. The 'No Entry' sign refers to the surrounding land, not to the access to the foreshore.

Nearest postcode: GL13 9HU (Berkeley Arms)

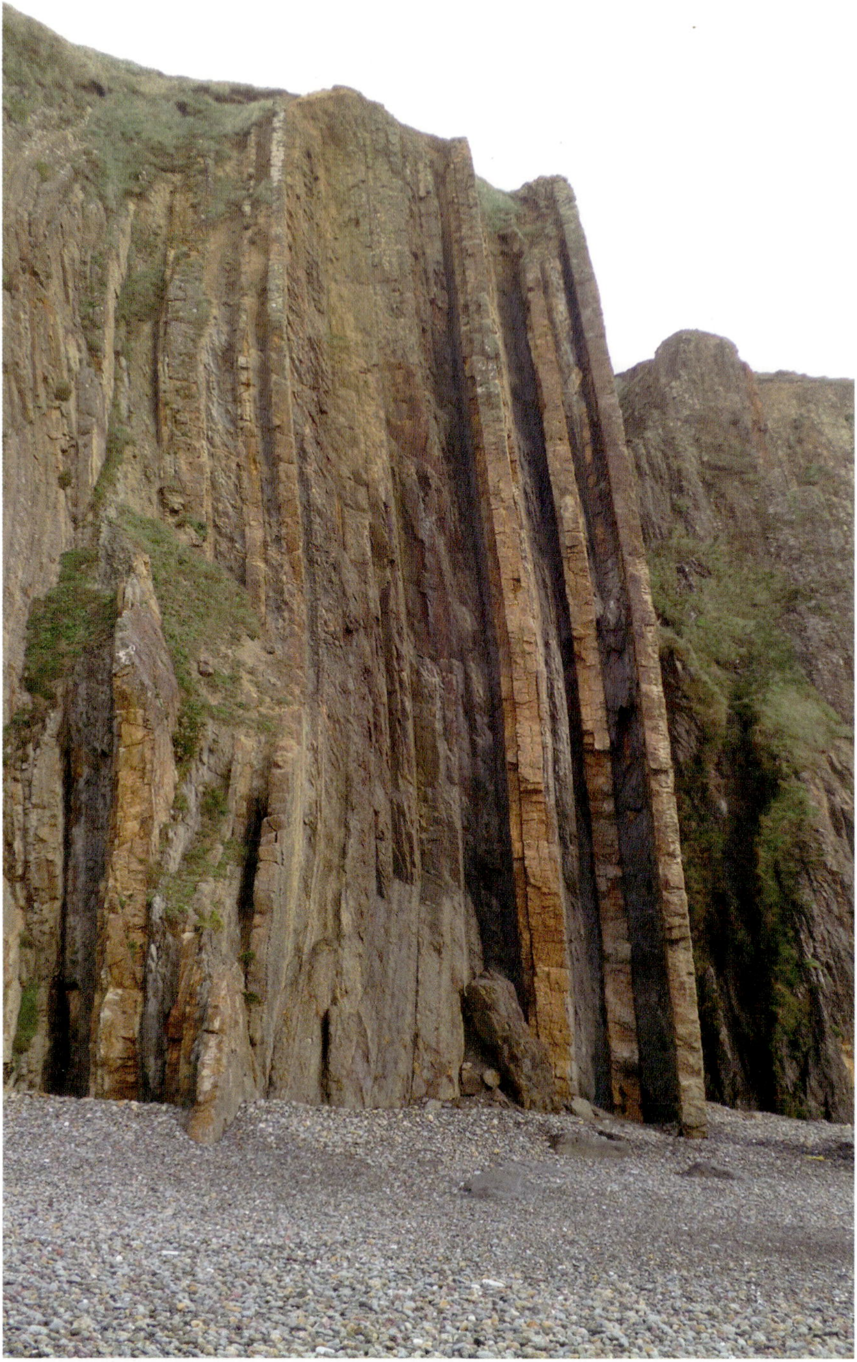

MARLOES SANDS, PEMBROKESHIRE

SSSI site – no hammering or digging into the cliffs

The Silurian rocks of Marloes Sands comprise those of the Aeronian (c. 440.8–438.5 mya), Sheinwoodian (c. 433.4–430.5 mya), Homerian (430.5–427.4 mya) and Gordtian (427.4–425.6 mya) stages, all of marine origin. The most fossiliferous of these rocks are those of the Coraliferous Group, from which fossils are generally obtainable from the scree and rocks on the beach, although erosional rates here are not high.

The rocks in the bay are uplifted and faulted, with the oldest to the north and youngest in the south. Fossils are often distorted or crushed, the most common being brachiopods, molluscs and coral.

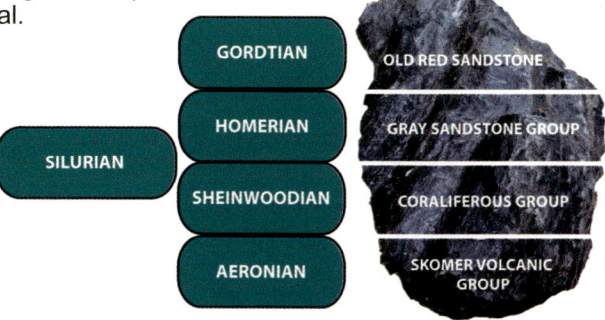

SILURIAN		
	GORDTIAN	OLD RED SANDSTONE
	HOMERIAN	GRAY SANDSTONE GROUP
	SHEINWOODIAN	CORALIFEROUS GROUP
	AERONIAN	SKOMER VOLCANIC GROUP

Marloes Sands is situated at Marloes peninsula, at the western edge of Pembrokeshire and located south west from Marloes village.

Fossil collecting at Marloes Sands may not be too productive. It is dependent on finding the best spot within the cliff face and scree below. Nonetheless, the Coraliferous Group should yield some brachiopods, bivalves or corals with some patient searching. Trilobites are found at this location but are much rarer. Corals are probably the most common fossil found here.

Marloes Sands is an SSSI site and along with much of Pembrokeshire, fossil collecting must be conducted by obtaining specimens from the scree. There is strictly no removal through hammering of the cliff face.

Collection requires little in the way of equipment, although a hammer might be useful to break larger rocks found on the beach, to save the weight being carried back to the car!

The beach provides a safe area of sand, which extends for approximately 1.5 km. Please be very aware of tides at this location and always consult a tide times website before visiting.

Site summary

⚠️ Rock falls

⚠️ Slippery rocks

⭐⭐⭐⭐⭐ Accessibility

⭐⭐⭐⭐⭐ Child suitability

⭐⭐ Find frequency

SSSI Site.
No hammering or digging into the cliffs.

Be aware of the tides.

Suggested equipment: Geological hammer for fallen larger rocks, rucksack.

Directions: 8 miles west of Milford Haven. Approach by way of the B4327 road to car park (pay). Access to the beach itself is via rough steps and across boulders.

Nearest postcode: SA62 3BH National Trust car park is half a mile from the beach.

Silurian rocks at Marloes Sands, containing various molluscs, gastropods and brachiopods.

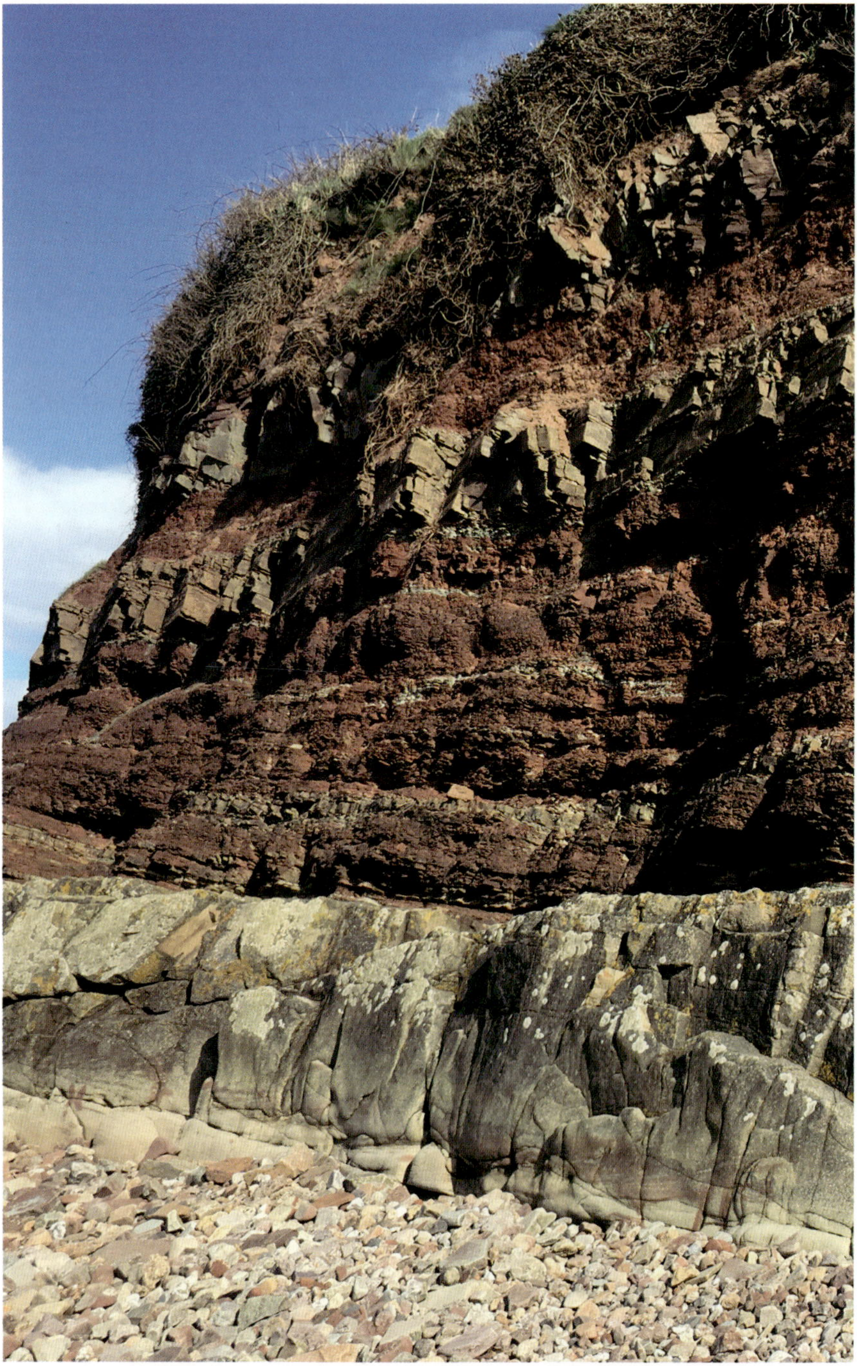

PORTISHEAD, NORTH SOMERSET

SSSI site – no hammering or digging in the cliffs or bedrock

Portishead is a coastal town on the Severn Estuary, close to Bristol and within the unitary authority of North Somerset. The rocks of Portishead date back to the Upper Palaeozoic (570–400 mya). The rocks mainly belong to the Old Red Sandstone (Fammenian) facies of the Upper Devonian (372.2–358.9 mya) that yield several important species of fossil fish and the Tournaisian stage of the Carboniferous, all clearly exposed in these coastal sections.

The most common fossils found in this area are fish scales and plates, burrows and crinoids, and if you are lucky, complete fishes and maybe even eurypterids (giant sea scorpions).

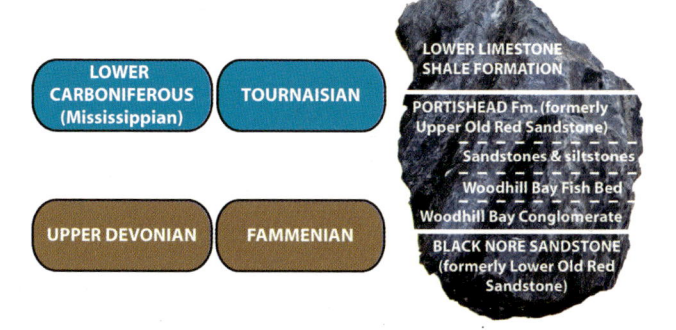

LOWER CARBONIFEROUS (Mississippian) — TOURNAISIAN

UPPER DEVONIAN — FAMMENIAN

LOWER LIMESTONE SHALE FORMATION
PORTISHEAD Fm. (formerly Upper Old Red Sandstone)
Sandstones & siltstones
Woodhill Bay Fish Bed
Woodhill Bay Conglomerate
BLACK NORE SANDSTONE (formerly Lower Old Red Sandstone)

The foreshore at Portishead offers an opportunity to walk through a sedimentary succession from the Lower Carboniferous through to the Mid-Lower Devonian. There is excellent exposure of the Portishead beds, the rocks that form the core of the Mendips, and the area has been known for many years for the famous Woodhill fish beds.

At Battery cliff, the Lower Limestone Shales are from the early Carboniferous and consist of alternating limestones and siltstones. The limestones contain skeletal debris and well preserved fossils. The transition from Carboniferous to Devonian cannot be seen as there is a shallow valley to the south of Battery Point which is infilled with Triassic sediments. The limestone units in Woodhill Bay are made up of a mixture of highly abraded skeletal debris and well preserved fossils of articulated crinoid stems and examples of the coral *Vaughania* (*Cleistopora*) which is the index fossil for the Lower Limestone Shales.

At Kilkenny Bay, at the Southern end of the sea wall, cliffs of the Upper Old Red Sandstone (including the Woodhill Bay fish beds) are well exposed. This is the Portishead Formation. This means that the rocks get progressively older to the South, eventually changing into the Black Nore Sandstones of the Lower Old Red Sandstones.

The Woodhill Bay fish bed formation consists of sandy siltstones and sandstones. Many fish have been discovered here in the past and fish scales can still be found.

Silurian fish fragments in beach pebble.

Fish scales and fragments from Devonian armoured fish can still be found at this location.

Crinoids in matrix from Carboniferous rocks at Battery Point, Portishead.

Site summary

⚠️ Slippery rocks

★★★★★ Accessibility

★★★★★ Child suitability

★★★ Find frequency

SSSI site. No hammering or digging in the cliffs or bedrock.

Best accessed on a falling tide as tides can reach base of cliff. Hard hats should be worn near cliffs. Slippery rocks in tidal areas, steep slopes on Battery Point. Strong tidal flows.

Suggested equipment: Chisel pick or geological hammer and bolster or splitting chisels, safety goggles, wrapping material, specimen bag.

Directions: Park on esplanade near Lido (all public access) or reach the site from Portishead town centre and take the South Avenue then South Road, to the junction with Leigh View Road. There, turn right to park near the Royal Hotel. Steps nearby descend to the pebble beach to the start of the section, beginning with the Carboniferous Pennant Sandstone near the pier, working west to Woodhill and Kilkenny Bay.

Postcode: BS20 7HG

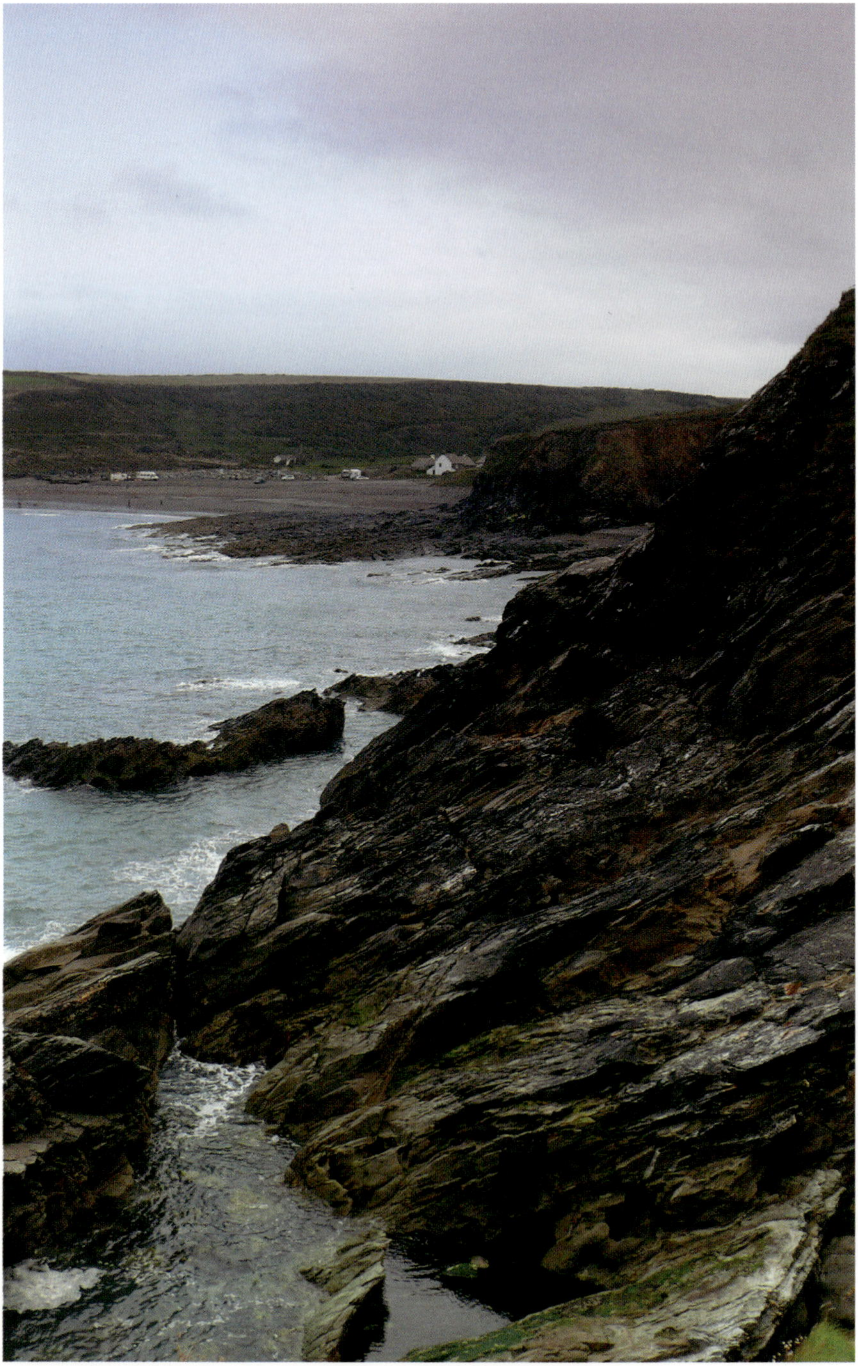

DRUIDSTON HAVEN, HAVERFORDWEST, PEMBROKESHIRE

SSSI site – no hammering or digging into the cliffs

Druidston is a long, sandy beach enclosed by steep cliffs, composed of Ordovician mudstone shales from the Mydrim Shales Formation (461–451 mya). These rocks are highly productive for the collection of graptolites found in the black shales.

The Pennant Sandstone Formation (310–307 mya) is also present here. This is a sub-unit of the newly established Warwickshire Group, recognised as Kasimovian to Asturian age (previously part of the Upper Coal Measures) from which the collection of Carboniferous plant fossils is possible. It is the only section of Upper Westphalian (Carboniferous) rocks in the western area of the South Wales Coalfield. Collection will be limited to fossils found on the foreshore in fallen rocks but this can still be productive, with patience.

CARBONIFEROUS — KASIMOVIAN/ASTURIAN — WARWICKSHIRE GROUP — PENNANT SANDSTONE FORMATION

ORDOVICIAN — CARADOCIAN — DREFACH GROUP — MYRDRIM SHALES FORMATION

The dark Ordovician shales at Druidston have been contorted by complex folds and faults and lifted above the Coal Measures rocks.

The rocks in Druidston Bay show unique, well-exposed Ordovician rocks, which demonstrate the effects of the Caledonian (400 mya) and Variscan (290 mya) mountain building periods. As a result, the rocks are folded, faulted and distorted.

Many of the graptolite fossils found here suffer from the same fate and can be disappointingly broken, especially specimens which do not follow the natural bedding planes of the shale rock.

The dark shale rocks and cliffs at the back of Druidston Beach.

The rocky southern end of Druidston Haven.

The Carboniferous plant fossils can be poorly preserved but certainly worthy of investigation among the sandstone rocks found here.

Site summary

⚠️ Infrequent rock falls

⚠️ Slippery rocks when wet

★★★★ Accessibility

★★★ Child suitability

★★★★ Find frequency

SSSI site.
No collection of large rocks or boulders, no hammering.

Suggested equipment: Specimen bags.

Directions and parking:
Access to the beach is by the Pembrokeshire Coastal Path or two footpaths joining up with the lane that runs from Nolton Haven to Broad Haven. The road to The Druidstone Hotel is the nearest route by road.

Nearest postcode (The Druidstone Hotel): SA62 3NE

Graptolite fossils found at Druidston Haven.

CAIM, PENMON, NORTH WALES

SSSI site – no hammering or digging into the cliffs

The Caim peninsula is a remote section of the North Wales coastline, near Penmon in Angelsey. The Carboniferous Limestone Series (of the mid-Mississippian) found here is of Visean age, and part of the Clwyd Limestone Group. It is dated between 346.7–330.9 mya and the cliffs show a range of limestone facies, with sandstone and mudstone units. Characteristic of the coastline here, some dolomitisation is present (a process by which dolomite is formed, when magnesium ions replace calcium ions in calcite) but the fossils include beautiful fossil corals in situ in the cliffs and as wave worn pebbles on the beach. Horn corals can also be found, along with silicified brachiopods and bivalves.

Caim is not an easy place to find, despite the optimistic guides online! It's also a difficult beach to get down to, after a long walk.

CARBONIFEROUS   VISEAN   CARBONIFEROUS LIMESTONE   CLWYD LIMESTONE GROUP

The beach at Caim, looking north and showing the rocky foreshore.

The headland at Caim.

A piece of coral found among boulders at Caim.

The rocks at Caim are renowned for their coral colonies, many of which have been weathered, to stand in relief from the limestone matrix in which they are embedded. You should not attempt to collect in situ fossils but photographs of the many varieties are recommended. The corals also appear in the many waveworn pebbles along the beach and occasional pieces of the Carboniferous limestone, with corals, can be found among the rocks and larger boulders on the foreshore.

The find frequency here for fossil colletion is low. The fossil corals are plentiful, but are mainly confined to the limestone cliffs, which are worthy of exploration. Photography of the in situ specimens is advised.

The corals stand proud of the weathered matrix of the Carboniferous Limestone bedrock and are easy to observe. The pebbles and cobbles that make up the beach also contain fossils. Brachiopod sections and pretty corals are frequently found and with some searching you should be able to collect quite a few of these, as wave worn pebbles.

The coastal scenery around Caim is spectacular, but this is a remote location with complicated parking and access, entailing some difficult walking. Ensure you have a mobile phone and have notified others of your intent to visit this site and when you are likely to return.

Coral fragment found among boulders at Caim foreshore.

Coral found in situ within the cliffs at Caim.

Site summary

⚠️ Infrequent rock falls

⚠️ Slippery rocks when wet

★★ Accessibility

★ Child suitability

★★ Find frequency

SSSI site.
No hammering in cliff face.

Suggested equipment: Specimen bags. Camera.

Directions and parking: You are able to park on a small concrete layby about a mile from Caim (on the approach road), but this is not for anyone with mobility issues or restricted walking ability.

Alternatively, head to Penmon, Beaumaris, Gwynedd, Wales, LL58 8RW
Here, you will get to a toll, which will cost £2.50 for the whole day, including lighthouse access, parking and access to the beachfront. From here, you are able to take the cliff trail walk to Caim.

Postcode: LL58 8RW

SEAHAM, CO. DURHAM

SSSI site – no hammering or digging into the cliffs

The south-eastward dipping rocks at Seaham are the highest Permian strata exposed on the Durham coast. The sequence exposed comprises the uppermost part of the Roker (Dolomite) Formation (8+ metres), the Seaham Residue (up to 9 metres) and, at the top, the Seaham Formation (about 31 metres). Fossils abound in the Seaham Formation and comprise two species of bivalve and the supposed alga *Calcinema permiana*, displayed as tiny stick-like tubular remains and present in enormous numbers in much of the rock.

Seaham also provides an Upper Carboniferous Coal Measures spoil heap at Nose's Point, which was dumped in front of the Seaham Formation cliffs from the old Dawdon Colliery. The colliery closed in 1991, but the tall cliffs of spoil continue to yield well preserved plant fossils.

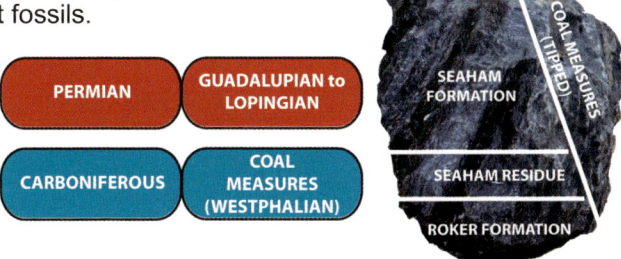

PERMIAN — GUADALUPIAN to LOPINGIAN

CARBONIFEROUS — COAL MEASURES (WESTPHALIAN)

COAL MEASURES (TIPPED)
SEAHAM FORMATION
SEAHAM RESIDUE
ROKER FORMATION

This coastal site is the type locality and by far the best surface exposure of both the Seaham Formation and the Seaham Residue. It is also the best surface exposure of the highest beds of the Roker Dolomite Formation.

The geology of the Durham coast is characterised by the extensive cliff and foreshore exposures of the dolomites and limestones of later Permian age. These rocks (formerly known as the Magnesian Limestone), were deposited in the Zechstein Sea in a relatively shallow, landlocked sea that extended from northeast England to Poland.

This is an SSSI Site, so please be aware that no hammering or digging into the cliffs is allowed, although Permian molluscs may be collected from fallen blocks.

The tipped slag heaps from the former colliery, on the other hand, can be dug through and from which a very good number of plant specimens of Carboniferous age can be collected. The slag heap is gradually being washed away by the sea but whist it is still there, it is a very productive site.

The best method for fossil collection is to split the rocks found on the foreshore or within the spoil heap. The rocks can produce a number of plant species, including *Neuropteris*, *Lepidodendron* and *Stigillaria*.

The fast eroding spoil tip from the Coal Measures, which forms part of the cliff face at Seaham and obscures the Permian strata beneath.

Bark of *Lepidodendron*, also known as the scale tree.

The fern *Neuropteris*, in a nodule from the Coal Measures spoil tip.

The alga, *Calcinema permiana*, is displayed as tiny stick-like tubular remains within the Permian rocks found along the coast here.

Site summary

Infrequent rock falls

Slippery rocks when wet

★★★★ Accessibility

★★★★ Child suitability

★★★★ Find frequency

SSSI site.
No hammering or digging in the cliffs or bedrock.

Steep access down slipway to beach.

Suggested equipment: Geological hammer, chisel, wrapping materials, specimen bags.

Directions: In Seaham, follow the A182 along the promenade and take the most southern roundabout, leading to a car park. From here, walk towards the fossil tree sign and down the slipway (which forms part of the spoil).

Postcode: Terrace Green Car Park, North Terrace, Seaham SR7 7EU

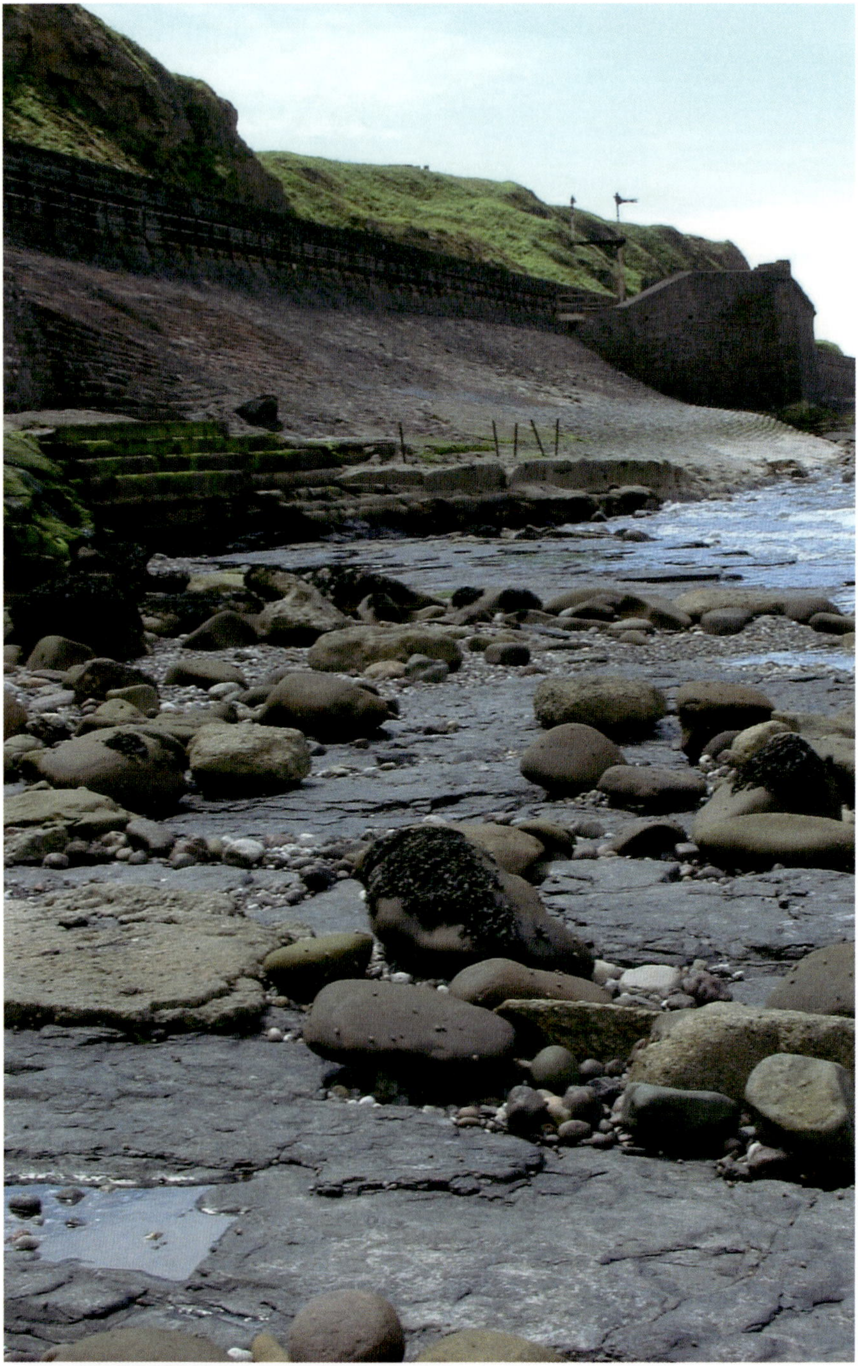

WHITEHAVEN, CUMBRIA

Access is permitted but no hammering in cliff face

The foreshore and cliffs at Whitehaven are famed for their Carboniferous plant remains. Many of the plant fossils that can be obtained here are of exceptional preservation and whilst the section in the cliff provides good collecting opportunities, the section of foreshore beneath exposes beds of Kasimovian age (311.7–306.5 mya, formerly Westphalian C) and generally consists of far better fossil material. It is also a safer option, as the cliff section is very overgrown in places and the scree slopes are potentially hazardous and difficult to traverse.

Safe access is through a tunnel, which leads directly onto the foreshore, rather than crossing the rail track and negotiating the sea defences. Take the A595 before you get into Whitehaven and turn off on Parton Brow. Drive to the bottom of the hill, to a waste ground car park at the bottom at CA28 6NY. The tunnel under the railway is only 10 metres from the car park. The site is very slippery, with a rock-strewn foreshore and is unsuitable for young children.

CARBONIFEROUS   KASIMOVIAN

WHITEHAVEN SANDSTONE SERIES

WHITEHAVEN SILTSTONES

COUNTESS SANDSTONES

At Whitehaven, the cliff is initially largely unfossiliferous but further northwards, particularly at Parton Cliff, the fossil-bearing shales emerge and can be split easily enough, using a geological hammer and chisel to reveal the plant material within. Safety goggles or similar should be worn as eye protection.

Collection on the tall and constantly crumbling cliff will involve climbing the scree slopes, which can be slippery and hazardous. The site is certainly not recommended for families with children, especially on the cliff and scree slopes. The foreshore rocks can also be slippery, especially when wet. *Calamites* stems and roots, some very large, are common finds.

The various rock layers can be clearly seen in the cliff and collectors will find that the shale with the brown or grey coloration are more productive for the plant fossils found here. Ascertaining from which particular layer the fossil was derived is certainly more problematic amongst material picked up on the scree slope or foreshore, but all fossil plants found at this location are of Kasimovian age, being 307–303.7 mya. They represent a time when plant life flourished and forests were populated by giant cycad trees and ferns. A substantial river once flowed to the southwest through this environment and the Countess Sandstone provides us with evidence of this.

The foreshore provides safer and better collecting and more than 30 different types of plant are known from this location. The clearly visible roots of the giant tree-like horsetail *Calamites* (shown above) can be found in abundance on the foreshore and in the lower face of the cliff. Splitting the shale between the roots will reveal further plant material, including *Neuropteris* (a seed fern), *Annularia* and *Asterophyllites* (horsetails). Many plants here can be found in their original life positions, making this location scientifically important.

Calamites stem on the foreshore.

Plant stem with a thorn.

Neuropteris, a seed fern from the foreshore shale.

Site summary

⚠️ Rock falls

⚠️ Slippery rocks

★★★★ Accessibility

★ Child suitability

★★★★★ Find frequency

Crumbling cliffs and rocky foreshore make this an unsuitable site for children. Safe access to foreshore through tunnel.

Suggested equipment: Chisel pick or geological hammer and bolster or splitting chisels, safety goggles.

Directions: Take the A595, before you get into Whitehaven turn off on Parton Brow, drive to the bottom of the hill and there is a waste ground car park at the bottom. You will see the tunnel under the railway (it is 10 metres from the car park).

Postcode: CA28 6NY

Continuing north to Parton Cliff, siltstones can be found in the shale. These contain ironstone nodules, which when broken may contain well preserved plant fossils.

BETTESHANGER COUNTRY PARK, DEAL, KENT

Betteshanger (formerly known as Fowlmead) Country Park is an easily accessible site, with easy parking and a visitor centre with toilets and a cafe. It forms 200 acres of park which is the former site of Betteshanger Colliery, Kent's last coalfield, which closed in 1989. Fossil plants are a common find here and the frequency of finding some nice specimens is very high.

Rocks to be found here are Upper Carboniferous in age, from the Bashkirian (323.2–315.2 mya) to Moscovian (315.2–307 mya). The rocks are from between 316–311 mya and mostly consist of the Kent 5 coal seam, with some Kent 7 scattered under the bench. Kent 5 is assigned to the Upper Coal Measures (Warwickshire Group) and Kent 7 to the (South Wales) Middle Coal Measures.

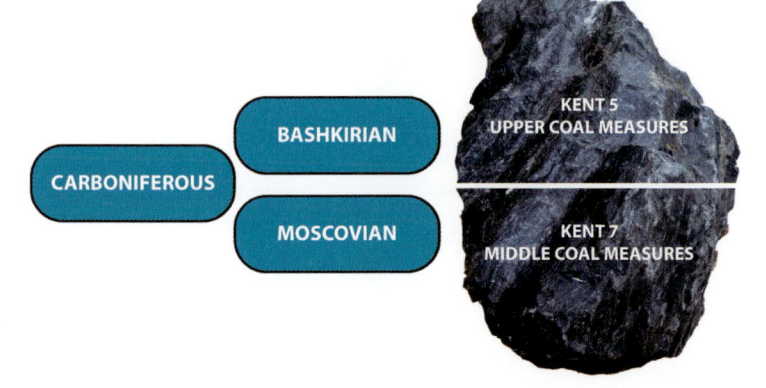

The former Betteshanger Colliery spoil tip was regenerated into Fowlmead Country Park, which was renamed Betteshanger Country Park in 2015. Geoconservation Kent have retained a section of the spoil tip for fossil collection purposes and groups can be accommodated on the site, by prior permission.

Fossils at the spoil tip are best found by digging into the bank and finding the hard pieces of shale rock within it. A spade is best for this work, although a trowel is adequate. The pieces of hard shale are plentiful but require splitting. Split the rock, if possible along the bedding plane, using a geological hammer and chisel. A flat bolster chisel is often ideal. Protective goggles are essential.

Also, be prepared to get very dirty. This is basically a spoil tip of the former coal mine and coal dust and grime are par for the course. The fossil plants found here are plentiful and can often be quite spectacular.

Wrap any finds in newspaper or bubble wrap for transporting. All fossil plants can be kept, but any insects and especially the large arthropod *Arthropleura* (which has been found at the site) must be reported to the visitor centre.

The different fossil plant assemblages already found at Betteshanger Country Park indicate areas of dense forest, river levees and overbank (crevasse) deposition during Carboniferous times.

Neuropteris leaf.

Alethopteris leaves.

Section of *Stigillaria* stem.

Site summary

★★★★★ Accessibility
★★★★★ Child suitability
★★★★★ Find frequency

A safe site, with plenty of good specimens to be found.

Toilets, cafe & car parking within a short distance of the fossil area.

Suggested equipment: Chisel pick or geological hammer and bolster or splitting chisels, safety goggles.

Limited access as permission is required.

Directions: Betteshanger Country Park is situated off the A258, at Sholden, near Deal, Kent.

Postcode: CT14 0BF

For bookings email: bcpinfo@betteshangerparks.co.uk
Telephone number: 07824 569181

NEWHEY QUARRY
ROCHDALE, LANCASHIRE

Newhey Quarry is a long abandoned quarry, well known for its Carboniferous plant material and marine fauna.

The massive beds of Lower Coal Measures sandstone – the Milnrow Sandstone at Newhey – was the reason for the quarry's existence and had been used extensively in the area for building use. Beneath this thick, prominent bed, a thin bed of mudstones and siltstones overlay a marine bed with bivalves (such as *Carbonicola* sp.) and from which ripple marks can be found in a thin bed at the base.

The Milnrow Sandstone is buff in colour and darkens to ocherous. It is overlaid by a thick seam of coal. The rocks are Bashkirian in age, from the Upper Carboniferous (Pennsylvanian) and are from 316–306 mya.

CARBONIFEROUS   BASHKIRIAN

COAL MEASURES
MILNROW SANDSTONE
MUDSTONES & SILTSTONES
MARINE BED
with ripple marks

Situated east of Rochdale, near the Junction 61 exit of the M62, the quarry can be seen clearly off the A640 near New Hey station and parking is straightforward enough.

The plants of Newhey fall into three categories; fern pinnules, *Calamites* and *Stigmaria*. Some sources refer to fossil fish and insects here but they are extremely rare and few, if any, have been found in recent years. Fossils are derived from the Milnrow Sandstone, which constitutes thin beds separated by layers of white mica, varying to coarse massive cross-bedded grit.

The best fossils are found in hard nodules that periodically fall out of the cliff face. The pinnules tend to be in smaller, roundish nodules that are reddish orange in colour; whilst the *Calamites* are in larger, harder greyish nodules that resemble elliptical cylinders. The *Stigmaria* are often found nodule free, as they fall out of their surrounding matrix when they hit the ground.

The site is replenished frequently from rock falls, which collectors need to be aware of. We would certainly recommend the wearing of hard hats at this site. The site is best for older children, due to the hazard of falling rocks and also the amount of standing water to be encountered.

Our grateful thanks to Andrew Eaves for providing the photos and much of the information on New Hey Quarry.

Section through the Milnrow Sandstone.

Lycopods *Sigillaria* and *Lepidodendropsis*.

Calamites growth tip, from a Milnrow Sandstone nodule.

Example of the Marine Bed's ripple marks.

Site summary

⚠️ Rock falls

⚠️ Slippery rocks

★★★★★ Accessibility

★★ Child suitability

★★★★ Find frequency

Exercise care at this old quarry site. Rocks are stable but still fall from above.

More suited to older children.

Suggested equipment: Chisel pick or geological hammer and bolster or splitting chisels, safety goggles, wrapping material (newspaper or bubble wrap).

Directions: Turn off the main road, behind St. Thomas' School and the Lower Bird in the Hand pub. Newhey Quarry, 41 Huddersfield Road, Newhey, Rochdale.

Postcode: OL16 3QZ

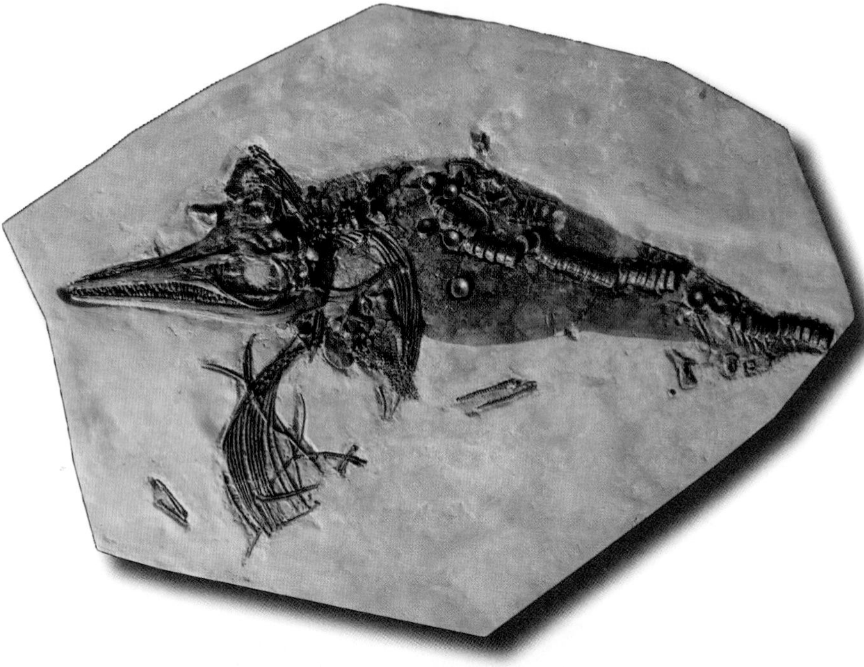

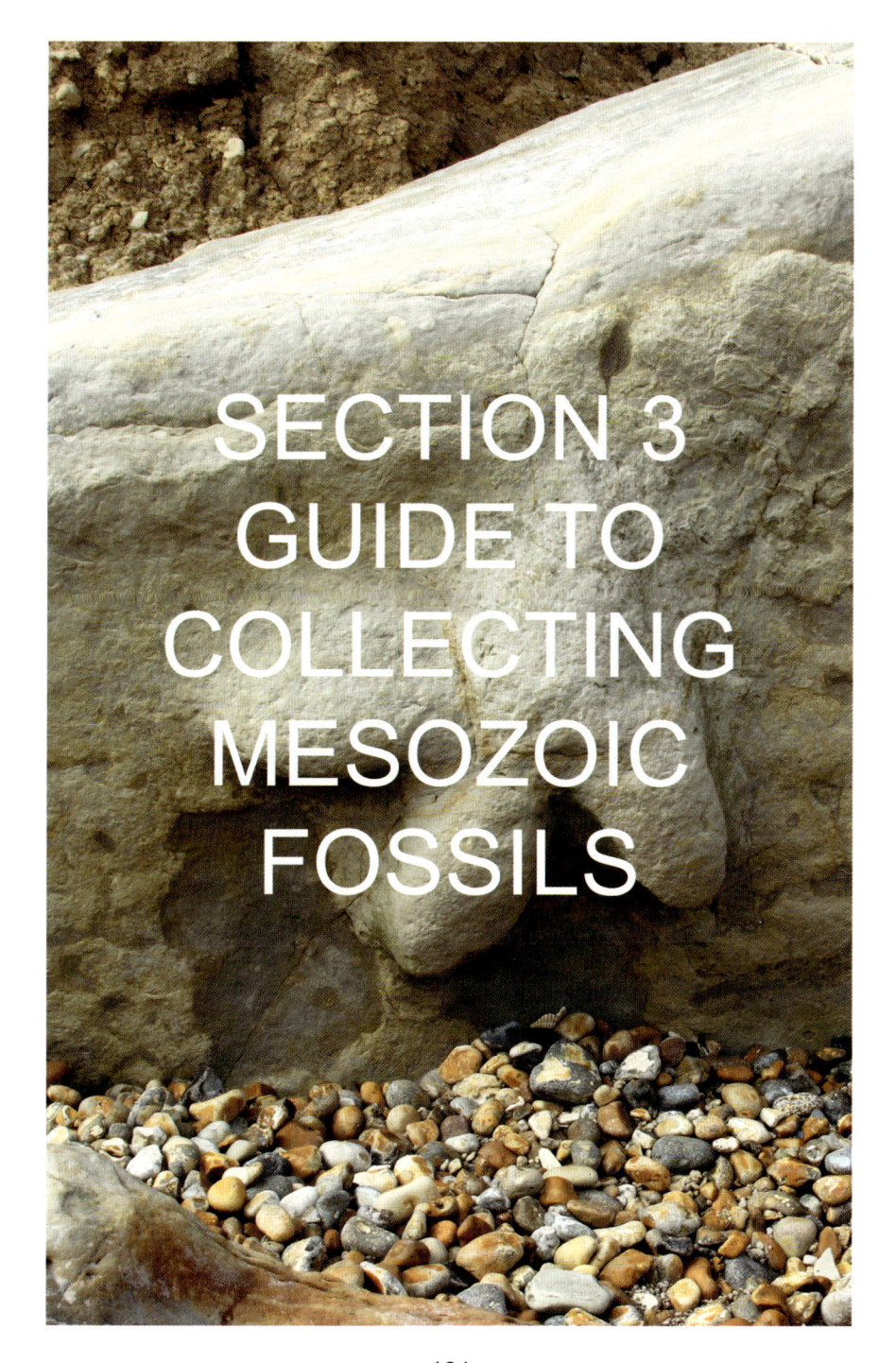

SECTION 3
GUIDE TO COLLECTING MESOZOIC FOSSILS

MESOZOIC FOSSILS

The British rocks of the Mesozoic era occupy a large area of the UK and offer the fossil collector many sites, with a varied, extensive range of flora and fauna. This diversity of fossils is due to the many different environments that existed during the 186 million years of the Mesozoic era, which is divided into three periods of geological time; the Triassic, Jurassic and Cretaceous.

During the Triassic period, arid desert conditions existed, which gradually changed as Britain slowly drifted northwards on the Eurasian plate and during the break up of the 'super continent' of Pangea, during the Jurassic period that followed. The Jurassic saw much of Britain underwater again, with sedimentary rocks being deposited. Sea levels rose and fell, creating environments for marine life, as well as for the dinosaurs, to evolve and flourish. The Cretaceous period witnessed the formation of the modern continents and of the Atlantic Ocean. Britain was also submerged again, as sea levels rose and vast deposits of Chalk were laid down.

Fossil collectors can be spoilt for choice in obtaining fossils from Mesozoic sites but the locations that we have chosen should, given the right conditions, provide a good opportunity for collecting some very good specimens for your collection and allow you to examine some classic Mesozoic exposures across the UK.

MESOZOIC FOSSIL SITES

The following Mesozoic sites are described in this section, with the following colour code applied, as shown in the geological time chart below.

Triassic
Sidmouth, Devon
Aust, Gloucestershire

Jurassic
Burton Bradstock, Dorset
Seatown, Dorset
Ringstead Bay, Dorset
Monmouth Beach & Church Cliffs, Lyme Regis, Dorset
Runswick Bay, North Yorkshire
Lavernock, South Wales
Watchet, Somerset
King's Dyke, Wittlesea, Cambridgeshire
Hampton Vale Lake, Yaxley, Cambridgeshire
Charmouth (Black Ven), Dorset
Charmouth (Stonebarrow & East Beach), Dorset
Saltwick Bay, North Yorkshire
Penarth, South Wales
Irchester Country Park, Northamptonshire
Quantoxhead, Somerset
Sedbury Cliff, Gloucestershire
Cross Hands Quarry, Worcestershire

Cretaceous
Folkestone, Kent
Danes Dyke, East Riding of Yorkshire
Compton Bay, Isle of Wight
Yaverland, Isle of Wight
Eastbourne, East Sussex
Hunstanton, Norfolk
Seaford, East Sussex
Pett Level, East Sussex

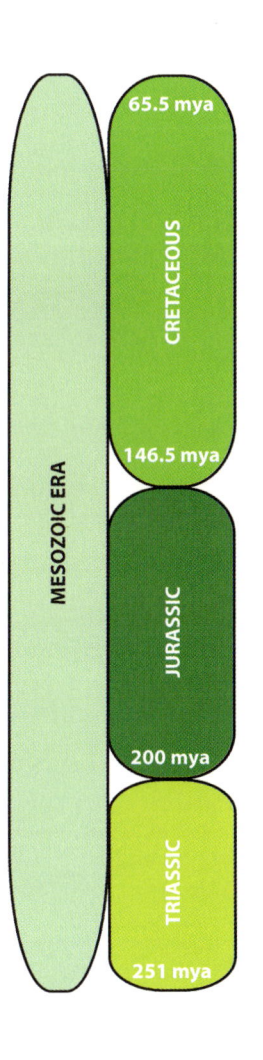

The Middle Triassic rocks at Ladram Bay, Sidmouth. The Triassic reptile jaw found by Chris Moore and the photo used with kind permission of Sidmouth Museum.

SIDMOUTH, EAST DEVON

SSSI Site – no hammering or digging into the cliffs

The cliffs at High Peak, are the eroded face of Peak Hill, directly to the west of Sidmouth town. This is a 118 metre-thick Triassic sequence of largely fluvial sandstones, including breccias and thin mudstones layers. At High Peak, the uppermost part of the cliffs are Cretaceous Upper Greensand, overlying the Triassic Sidmouth Mudstone Formation (formerly the Mercia Mudstone) in the centre of the cliff, with the Otter Sandstone Formation below.

The Triassic rocks displayed below High Peak have yielded very rare remains of Middle Triassic fish, amphibians and reptiles. Most specimens have been recovered from fallen blocks but a few have been found in situ. Bones and footprints of the labyrinthodont *Mastodonsaurus lavisi* and the rhynchosaur *Fodonyx spenceri* have also been found on the foreshore. Comparison of the fauna indicates an Anisian age for the Triassic Otter Sandstone Formation of 247.2–242 mya.

MIDDLE TRIASSIC   **ANISIAN**

UPPER GREENSAND

MERCIA MUDSTONE
(SIDMOUTH MUDSTONE Fm.)

OTTER SANDSTONE FORMATION

Fossil collecting at Sidmouth is not going to be productive to the casual collector, and in any case, must be confined solely to the loose blocks found on the beach due to the sensitivity of this scientifically important site. The pebble lags (the conglomerates generally confined to the basal part of a channel fill during sediment deposition) are the best place to look. It is a geologically important site and one of only a few Triassic sites in Britain. Fossils likely to be found are rhizoconcretions; the roots of Triassic plants. Most fossils found, including any bone, are usually only fragmentary.

The Triassic sediments at Sidmouth were laid down in a vast sandy desert floodplain, covered with small ephemeral braided streams and lakes, with a semi-arid climate, that allowed long dry periods punctuated by seasonal rains and flash-floods. The Otter Sandstone Formation was deposited on the bottom of this shallow, oxygen-rich river and was an environment far from ideal for the formation of fossils.

The freshwater of the temporary rivers and streams would have been home to fish and amphibians, with vegetation along the edges. When the water stopped flowing, the sun drew moisture out of the ground and in the process, minerals such as calcium cemented the ground around the roots, forming 'rhizoconcretions' – which can occasionally be found on the beach. The fauna is dominated by skeletal material of a group of medium-sized herbivorous reptiles, known as rhynchosaurs.

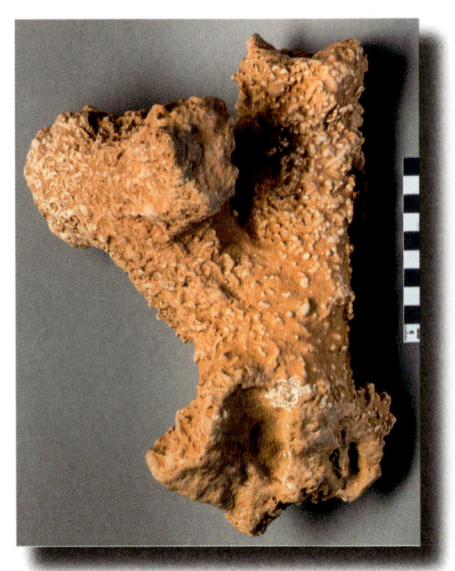

Rhizoconcretion found by Richard Edmonds. Photo used with kind permission of Fairlynch Museum, Budleigh Salterton.

Ripple marks found on the beach at Peak Hill. Photo used with kind permission of Sidmouth Museum.

Chirothere footprint from the foreshore below Peak Hill. Found by Rob Coram and housed in the Sidmouth Museum. Photograph used with their kind permission.

While in Sidmouth, be sure to visit the Sidmouth Museum in Church Street, where a number of Triassic fossils from the area are housed.

Site summary

⚠️ Rock falls

⚠️ Slippery rocks

★★★★ Accessibility

★ Child suitability

★ Find frequency

SSSI Site. No hammering or digging into the cliffs of this protected Jurassic Coast World Heritage Site.

The cliffs are prone to sudden collapse and falls are frequent. The rocky foreshore and tidal location makes this an unsuitable site for children. Confine any foreshore collecting to fallen blocks on the beach and especially the lag conglomerate deposits.

Suggested equipment: Specimen box, wrapping material.

Directions: Car parking directly opposite Chit Rocks along Peak Hill Road and in nearby car parks. Access to the site at low tide via coastal path. There is also car parking at the nearby campsite and caravan park at Ladram Bay, from where there is direct access down to the beach or out to the coastal path.

Postcode: EX10 0NW

The Middle and Upper Triassic rocks at Aust, overlaid by Lower Jurassic Blue Lias.

AUST, GLOUCESTERSHIRE

SSSI Site – no hammering or digging into the cliffs

This Middle and Late Triassic and Early Jurassic site lies south (downstream) of the Severn Bridge and can be seen as the familiar white and red cliffs. It is one of the UK's few productive Triassic fossil collecting localities.

Here, the red beds of the Mercia Mudstone Group (formerly called Keuper Marls) form the Branscombe Mudstone Formation (221–206 mya) from the cliff base, passing up to the green-grey beds of the Blue Anchor Formation (221–206 mya), (formerly called Tea Green Marls). Then rest the darker, then lighter, grey beds of the Penarth Group: the Westbury Formation and Cotham Member (formerly called Rhaetic Beds) from the Late Triassic or Rhaetian (210–206 mya). At the cliff top are the light brown beds of the Blue Lias Formation from the Early Jurassic.

LOWER JURASSIC	HETTANGIAN
UPPER TRIASSIC	PENARTH GROUP
MIDDLE TRIASSIC	MERCIA MUDSTONE GROUP

- BLUE LIAS FORMATION
- COTHAM MEMBER
- WESTBURY FORMATION
- Bone bed
- BLUE ANCHOR Fm.
- BRANSCOMBE MUDSTONE Fm.

The Severn foreshore.

Organised fossil hunt beneath Aust's red and white cliffs.

Accessibility to the cliffs from Junction1 of the M48 is by way of a steel gate (with a stile) to the concrete causeway. Parking is on the B4461 Aust Wharf Road at Old Passage. The site needs to be approached with some caution, as the cliffs here are dangerous, with falling rocks. Beware also of the tides and the mud flats along the foreshore.

The foreshore provides the best and safest area for collecting. Examine the loose blocks on the beach, paying particular attention to the shingle, in which shells, teeth and pieces of bone may have collected.

The Bone Bed, for which Aust is famous, is invariably harder to find, due to over-collection. Frequent cliff falls do bring down fresh material, however and this can be split on the beach and fragments removed for preparation at home. The Jurassic Lias above can also provide good finds.

The Branscombe Mudstone Formation at the cliff base is unfossiliferous. It is the beds of the Penarth Group: the Westbury Formation and Cotham Member, from which the Bone Bed emerges and is totally reliant on cliff falls.

Plesiosaur vertebrae from the Blue Lias Formation at Aust.

Shrimp burrows in rock from the Penarth Group on Aust's foreshore.

Typical Bone Bed material containing bone and teeth.

Site summary

Rock falls

Slippery rocks

★★★★ Accessibility

★★★ Child suitability

★★★★ Find frequency

SSSI Site. No hammering or digging into the cliffs.

Approach site with caution. Falling rocks from cliff. Beware of tides and the mud flats along the foreshore, and the Severn bore (a large wave that happens twice a year). Children need supervision.

Suggested equipment: Geological or lump hammer, chisels, goggles, hard hat, wrapping material.

Directions: Aust is located on the eastern side of the Severn estuary, close to the eastern end of the Severn Bridge. Accessibility to the cliffs from Junction 1 of the M48 is through a steel gate (with stile) to the concrete causeway. Parking is on the B4461 Aust Wharf Road at Old Passage. Park on Passage Road (Old Passage). Head along Passage Road towards the bridge, until you get to the beach.

Postcode: BS35 4BG

Fallen slab of Bridport Sands at Burton Cliff in April 2016. The cliff is eroding at an alarming rate and should be approached with extreme caution. Ammonites shown are *Parkinsonia parkinsoni* and *Homeoplanulites* sp.

BURTON BRADSTOCK, DORSET

SSSI Site – no hammering or digging into the cliffs

The highly unstable, jointed and fractured Burton Cliff at Hive Beach towards Freshwater and West Bay is not recommended for the casual collector. Frequent rock falls, especially during winter months, are dangerous and fatalities have occurred. However, when Burton Cliff falls, the layer of Inferior Oolite at the top of the cliff at the western end, will contain a haul of ammonites, nautiloids, gastropods, bivalves and other fossils that locals lie in wait for, often for years!

The attractive, sandstone cliffs mostly comprise the Bridport Sands Formation, a weak to moderately strong sandstone that becomes weaker when wet. It is capped by a 3 metre band of Inferior Oolite Formation (174–168 mya), a limestone which was formed in marine conditions not unlike the Caribbean seas of today. Above the oolite is a bed of Frome Clay (Fuller's Earth), a calcareous mudstone.

JURASSIC
- BATHONIAN
- BAJOCIAN
- AALENIAN
- TOARCIAN

FROME CLAY Fm. (FULLER'S EARTH)
INFERIOR OOLITE Fm.
BRIDPORT SANDS FORMATION

Weathering of the Bridport Sands in Burton Cliff.

Toarcian aged Bridport Sands in Burton Cliff.

A large fall at Burton Cliff in January 2014.

Just east of Burton Freshwater, the cliff is mostly of Bridport Sand (Toarcian, Lower Jurassic), although the top 2 metres are Aalenian (Middle Jurassic) with the ammonite *Leioceras opalinum*. Above is about a 7 metre-thick layer of Inferior Oolite (Bajocian).

Fossils are present in the Bridport Sand but are not overly abundant. Examine rocks and boulders on the beach (staying well away from the cliff base) where the ammonite *Leioceras opalinum*, belemnites and trace fossils (particularly *Thalassinoides*) can often be found.

The Bridport Sand formed at the front of a giant delta, that once fanned into the Jurassic sea from a landmass in the north. Its formation would explain the absence of bivalves and oysters as fossils within it. The bright yellow colour of the Bridport Sand is due to oxidation of their contained iron salts. Depending on the amount of calcite cement present in the sands, being deposited on the sea bed during Jurassic times, the sand layers alternate between hard and soft beds. These weather at different rates, resulting in the distinctive layers of the cliffs. There is also an important difference in the mica present in the softer layers, where mica flakes are distinctly crumpled and distorted, as opposed to the undistorted mica present in harder layers.

Blocks of fallen Inferior Oolite are easy to identify and are abundantly rich in fossils. Ammonites including *Parkinsonia dorsetensis*, *Strigoceras truellei*, *Garantiana garantiana*, *Bredia* sp. and belemnites, nautiloids, echinoids, shells (bivalves, gastropods and brachiopods) and sponges can be found in decent numbers.

The eastern end of Burton Cliff with ammonites (left to right) *Parkinsonia parkinsoni*, *Leptosphinctes* sp., *Parkinsonia dorsetensis* and the bivalve *Neocrassina modiolaris* from the Inferior Oolite.

Cliff falls at Burton Bradstock are common but invariably consist of the Bridport Sands Formation, a largely unfossiliferous rock. The photos were taken in January 2016.

Blocks of fallen Inferior Oolite may require the use of a heavy hammer and chisels in order to collect fossils. If you are lucky enough to arrive after a fresh fall, there is usually plenty to be found, so a large rucksack will be vital for carrying finds back to the car park. Unless you are local and visit often, you are going to have to work hard to make finds but there is often still a few specimens to be found, even weeks after a fall.

Although it is tempting to climb the falls, the cliffs are highly unstable after recent collapses and even small pieces of rock falling from a height can cause serious injury. You should search around the base of the fall and not the base of the cliff. If you can, carry any rocks you are working on away from the fall site and as far as possible away from the base of the cliff.

A fallen block of the Bridport Sands Formation, containing *Leioceras opalinum* ammonites.

A section of a *Parkinsonia* ammonite, found by Martin Curtis on the foreshore in 2016.

Site summary

Rock falls

★★★★ Accessibility

★ Child suitability

★★ Find frequency

SSSI Site. No hammering or digging into the cliffs of this protected Jurassic Coast World Heritage Site.

These cliffs are dangerous. It is strongly advised to keep away from the base of the cliffs.

Children will require close supervision. Easy access to foreshore from car park at Hive Beach.

Suggested equipment: Geological hammer, lump hammer and chisels, wrapping materials.

Directions: Access from the eastern end to Hive Beach, park at the National Trust car park (postcode: DT6 4RF) or on foot from the village, turning west at the Dove Inn, following the footpath, keeping east of the River Bride to the sea.

SEATOWN, DORSET

SSSI Site – no hammering or digging into the cliffs

Three miles east of Charmouth lies the coastal hamlet of Seatown. With its Early Jurassic cliffs and foreshore, Seatown is a highly fossiliferous site and best examined at beach level. This is an SSSI, so no hammering or digging in the cliffs is permitted.

The precipitous cliffs are highly unstable and the fossil collector is strongly advised to collect from the fallen blocks, which are representative of almost every horizon. West of Seatown, the Jurassic cliffs towards Golden Cap are mostly of Pliensbachian age (190.8–182.7 mya), consisting of clay and limestone in alternating layers, overlain by later Cretaceous rocks of Gault and Upper Greensand from the Albian stage.

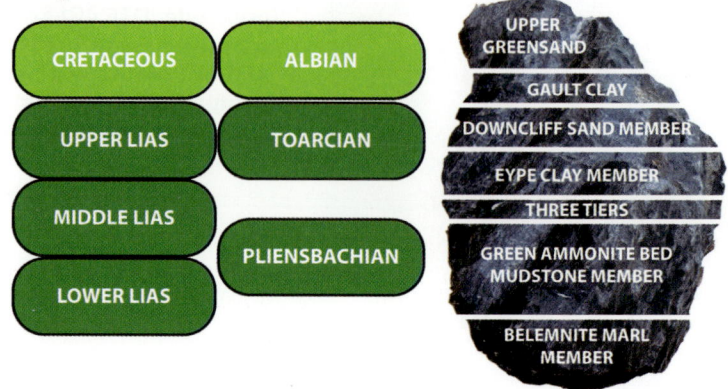

CRETACEOUS	ALBIAN
UPPER LIAS	TOARCIAN
MIDDLE LIAS	
LOWER LIAS	PLIENSBACHIAN

- UPPER GREENSAND
- GAULT CLAY
- DOWNCLIFF SAND MEMBER
- EYPE CLAY MEMBER
- THREE TIERS
- GREEN AMMONITE BED MUDSTONE MEMBER
- BELEMNITE MARL MEMBER

From the car park, walk west towards Golden Cap. After storms and the clearance of shingle, the uppermost part of the Belemnite Marl Member is often exposed, as a prominent layer of limestone with a profusion of belemnites. Above, lie the bluish-grey clays of the Green Ammonite Member. Both belong to the Early Jurassic rocks of the Lower Lias and are of Pliensbachian age. The upper part of the Green Ammonite Member is marked by the Three Tiers; calcareous sandstones, seen at a higher level. The uppermost part of the Green Ammonite Member is now considered to be Middle Lias and contains the ammonite *Amaltheus*.

Ammonites may be collected from fallen material, including from the Three Tiers. *Amaltheus* and *Tragophylloceras* are found in the Three Tiers, whilst *Androgynoceras lataecosta* is the 'Green' Ammonite'. Further ammonites such as *Aegoceras*, *Oistoceras* and *Liparoceras* can be found in the Green Ammonite Member, along with belemnites, bivalves and marine reptile bones. The limestone nodules found within this bed are the source of well preserved ammonites but require careful preparation.

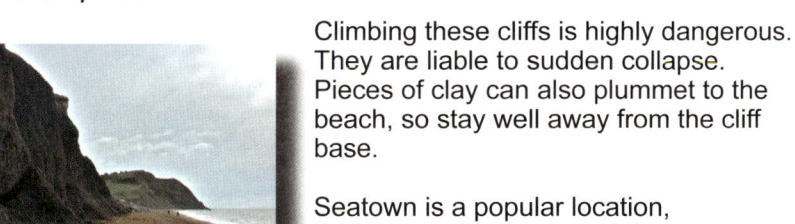

West towards Golden Cap and the ammonite *Liparoceras*.

Climbing these cliffs is highly dangerous. They are liable to sudden collapse. Pieces of clay can also plummet to the beach, so stay well away from the cliff base.

Seatown is a popular location, particularly during the summer months. The car park is busy, as is the pub and the beach can get overworked by enthusiastic fossil collectors. However, many prefer to stay close to the facilities and careful searching towards Golden Cap can often still be rewarding.

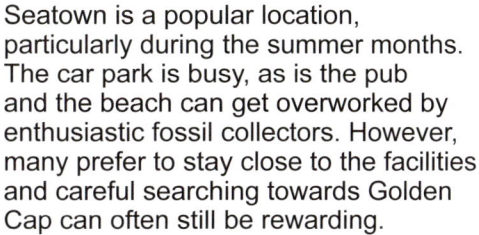

Green Ammonite Member in the cliffs at Seatown, looking east.

The section east of Seatown provides an opportunity to explore the Eype Clay Member, which reaches beach level due to faulting. The same faults have obscured the Green Ammonite Beds in this section. The Eype Clay Member is overlain by the Down Cliff Sand Member,

Ammonites from the Green Ammonite Member, including (clockwise from top left): *Lytoceras*, *Asteroceras obtusum*, *Asteroceras stellare* and *Tragophylloceras* (with *Androgynoceras* attached). Fossils kindly supplied for photographing by Tony Gill at Charmouth Fossil Shop.

followed by the Thorncombe Sand Member continuously east, as far as Eype. Ammonites, brachiopods and crinoids occur but the Down Cliff Sand is best known for its Starfish Bed, containing brittle stars, obtained from fallen blocks on the beach and are highly prized among collectors. If you do manage to find one, extraction is very difficult.

West of Seatown towards Charmouth, the grey clays of the Green Ammonite Mudstone member continue to Golden Cap.

Site summary

⚠️ Rock falls

⚠️ Slippery rocks

★★★★★ Accessibility

★★★★ Child suitability

★★★★ Find frequency

SSSI Site. No hammering or digging into the cliffs of this protected Jurassic Coast World Heritage Site.

Crumbling clay cliffs and tidal location. Children will require supervision. Easy access to foreshore from car park.

Suggested equipment: Geological hammer and chisels, small steel hand pick, specimen box, wrapping materials.

Directions: From Charmouth take the A35 to Chideock, turn right, opposite the church, into Duck Street and onwards down Sea Hill Lane to the coast. Parking is easy, with a large car park opposite The Anchor Inn. Toilets and refreshments available in the summer season.

Postcode: DT6 6JU

The western end of Ringstead Bay, at the headland of Bran Point, which consists of hard Corallian rocks from the Osmington Oolite Formation, with a wave cut platform, comprised of the bivalve *Myophorella clavellata*.

RINGSTEAD BAY, DORSET

SSSI Site – no hammering or digging into the cliffs

Ringstead Bay is situated east of Weymouth. Take the A353 out of Weymouth, through Osmington village and turn right before Poxwell village (signposted Ringstead Bay & Beach Cafe). Wind down the narrow lane until the road forks. Bear right onto the toll road and descend to the coast. You can easily park in the main car park (there is a day charge during the summer season). Toilets and a shop/cafe are next to the car park. From the car park, walk towards the beach, bearing right and then descend to the beach by way of a concrete slipway.

Ringstead Bay is a wonderful location, with rocks and fossils from the Corallian, Kimmeridge Clay, Purbeck and Portland Formation (and at the White Nothe headland, from the Upper Greensand and Chalk) all in one place!

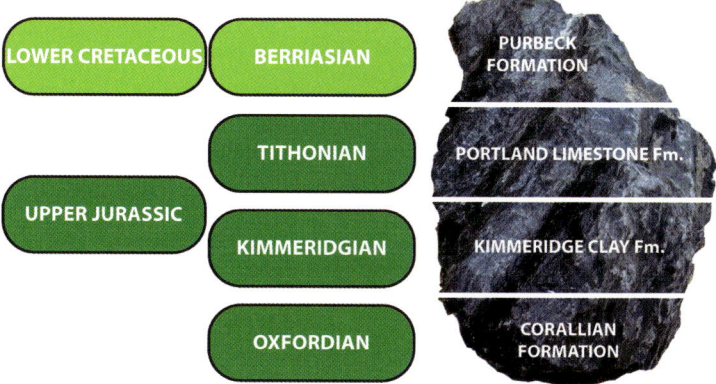

LOWER CRETACEOUS	BERRIASIAN	PURBECK FORMATION
UPPER JURASSIC	TITHONIAN	PORTLAND LIMESTONE Fm.
	KIMMERIDGIAN	KIMMERIDGE CLAY Fm.
	OXFORDIAN	CORALLIAN FORMATION

Turn right (west) and walk approximately 100 yards towards Bran Point, the nearest visible headland and just beyond the rock armour.

Here, the rocks are of Corallian age (the name of one of the divisions of the Jurassic rocks), from 155–150 mya. The first feature of interest along this section is the exposure of the Sandsfoot Formation, (previously known as the Ringstead Coral Bed). This marks the junction between the older Corallian rocks with the overlying, younger Kimmeridge Clay. The junction is marked by the occasional presence of the large, lopsided brachiopod *Torquirhynchia inconstans* (shown below) and is known as the Inconstans Bed.

Sandsfoot Formation below the Kimmeridge Clay, with brachiopod Torquirhychia inconstans.

At this location the Corallian consists of brown and dark green sandy clay, often with rich oolitic ironstone present. Occasionally, isolated corals can be found but in the main, colonial corals form part of a thin but distinguishable layer of limestone, protruding from the low cliffs at beach level.

The layer comprises corals, bivalves and other fauna from the warm Jurassic sea bed, which is also often found in broken rocks beneath the low, slumped cliffs at this section. Fossil collection can be made along the foreshore and not the cliffs.

On the foreshore and especially at low tide, can be seen the prominent Corallian rocks of the Osmington Oolite Formation and Clavellata Formation, consisting of the worn surfaces of thousands of thick shells within an incredibly hard limestone. This rock extends out into the sea as a wave-cut platform with distinct reefs.

Clavellata Formation, with the bivalve Myophorella clavellata.

The shells mostly belong to the large marine bivalve *Myophorella clavellata* (shown left). These shells are very difficult to extract and collection is best left to

The Kimmeridge Clay at Ringstead Bay and the gastropod Bathrotomaria reticulata.

The photo shows the Chalk dipping to the right, the Portland and Purbeck beds dipping to the left and Kimmeridge Clay in the lower part of the cliff.

Ammonite fragment from a fallen block of Portland Beds.

isolated specimens which can be found on the foreshore.

Turning back east and walking along the shingle beach towards the Chalk headland of White Nothe, the tall slumped cliffs of badly weathered Kimmeridge Clay crop out. The cliffs at the eastern end of the bay are known as Burning Cliff, due to a spontaneous combustion of the organic-rich bituminous oil shales within the clay. The fires burned from 1826 to 1830 but there has been no recent occurrence!

The Kimmeridge Clay is fossiliferous with many ammonites, bivalves and gastropods (especially *Bathrotomaria reticulata*, which locals call the 'Ringstead Snail' and *Bourgetia saemanni*). Worm tubes (*Serpula intestinalis*) are commonly found, along with the large oyster *Deltoideum delta*. Marine reptile remains are rarer but do occur.

High in the cliffs, east of Burning Cliff are the strata of the Portland Limestone Formation and the basal part of the Purbeck Formation. The rocks are folded and faulted, forming a spectacular unconformity, with the Chalk dipping to the right and the Portland and Purbeck Beds dipping to the left.

Fossils from both the Purbeck and Portland Limestone Formation often appear in slipped masses at beach level or in fallen blocks on the beach.

The fossil bivalve, *Ctenostreon proboscideum*. Partial specimens embedded in the rocks are a common find at Ringstead Bay.

White Nothe. The foreshore, beneath cliffs of Upper Greensand and Chalk, is strewn with rocks, some of which are septarian nodules with calcite.

Site summary

Infrequent rock falls

Slippery rocks

★★★★★ Accessibility
★★★★★ Child suitability
★★★★ Find frequency

SSSI Site. No hammering or digging into the cliffs of this protected Jurassic Coast World Heritage Site.

Crumbling clay cliffs. Do not climb. Rocky foreshore can be slippery. Child friendly beach. Easy parking and beach cafe with toilets.

Suggested equipment: Small hand pick, specimen box, wrapping material.

Directions: From Weymouth, follow the A353 east, through Osmington village and turn right (clearly signposted Ringstead/Beach Cafe). The narrow lane bends until a sharp junction. Turn right here, along the old toll road down to the bay and car park. From April until November there is a daily car park charge, which is very reasonable.

Postcode: DT2 8NG

MONMOUTH BEACH & CHURCH CLIFFS, LYME REGIS, DORSET

SSSI Site – no hammering or digging into the cliffs

To the west of the Cobb at Lyme Regis is Monmouth Beach and to the east, are Church Cliffs. Here the famous rocks of the Blue Lias Formation extend west to Pinhay Bay and to Black Ven in Charmouth to the east.

The distinctive cliffs of layered limestone and shale are present, particularly at Ware Cliffs. The Shales-with-Beef and Black Ven Marl Members overlay the Blue Lias and can appear as loose rocks and debris on the beach. Fossil collecting should only be attempted on the foreshore, as the cliffs are highly unstable and prone to collapse. The winter months are certainly the most productive time for finding specimens, which have fallen from the cliffs and have not been spotted by summer tourists!

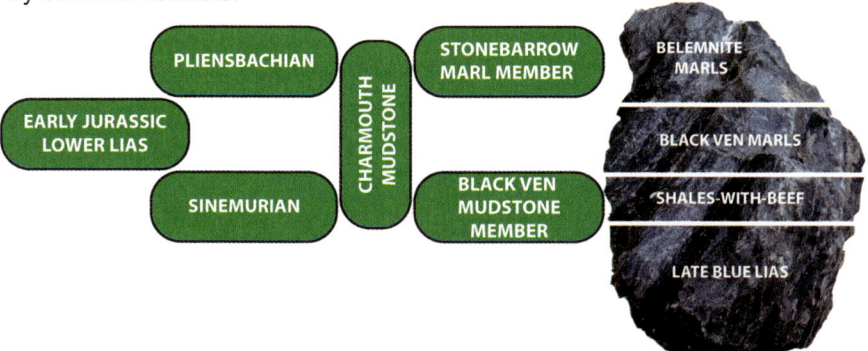

- EARLY JURASSIC LOWER LIAS
  - PLIENSBACHIAN
    - CHARMOUTH MUDSTONE
      - STONEBARROW MARL MEMBER
      - BLACK VEN MUDSTONE MEMBER
  - SINEMURIAN

Rock layers:
- BELEMNITE MARLS
- BLACK VEN MARLS
- SHALES-WITH-BEEF
- LATE BLUE LIAS

Church Cliff, Lyme Regis.

There is considerable risk from these cliffs and where possible, you should try and keep away from the beach at the foot of the cliffs, especially where fresh fallen material is evident. It is better to study the ledges, out on the shore at low tide.

The Blue Lias Formation consists of hard layers of limestones and shales and occupies much of the base level of cliffs around Lyme Regis and on the ledges, seen on the foreshore at low tide. East of Lyme Regis, Church Cliffs are a source of calcified ammonites. Above the Blue Lias, the Shales-with-Beef Member (part of the Charmouth Mudstone Formation) can be seen and consists largely of dark mudstones, with mostly crushed fossils. Above this lies the Black Ven Marl, which contains superb fossils, but at Lyme Church Cliffs the source is confined to sections that have fallen to the beach.

The whole stretch from Black Ven at Charmouth, to Monmouth Beach is rich in fossils. Ammonites, belemnites, crinoids, brachiopods, marine reptile bones and teeth, fish and molluscs regularly wash out of the clays along the top of the beach and with a keen eye can be picked up.

Searching for fossils can be the most productive if you look between the rocks, boulders, pebbles and rock pools. The issue at Lyme Regis is the keen competition from the thousands of people who visit, all trying to find a fossil. As a result, on some days there will be nothing to find, where the beach has been picked clean by others.

Promicroceras planicostum, a calcified ammonite from Church Cliff.

Walk as far west as you can on Monmouth beach at low tide and after some traversing of boulders, you reach a limestone ledge known as the Ammonite Pavement, where hundreds of ammonites of considerable size can be seen.

The ledges, rock pools and boulders are the most productive areas to look for fossils at low tide.

Plagiostoma giganteum, a large mollusc found here by Aidan Philpott.

Site summary

Rock falls

Slippery rocks

★★★★★★ Accessibility

★★★★★ Child suitability

★★★★ Find frequency

SSSI Site. No hammering or digging into the cliffs of this protected Jurassic Coast World Heritage Site.

Crumbling, unstable cliffs. Do not attempt to climb. Rocky foreshore can be slippery. Child friendly beach.

Suggested equipment: Geological hammer, chisels, goggles, specimen box, wrapping material. A hammer, chisel and eye protection are recommended for splitting prospective rocks, especially the nodules that might contain ammonites.

Directions: In Lyme Regis, park up at Holmbush car park (postcode: DT7 3HX), on the hill above the Cobb, walk down the steep hill of Cobb Road to the Cobb and then westward to Monmouth Beach. Parking is also possible at Monmouth Beach car park (postcode: DT7 3JN).

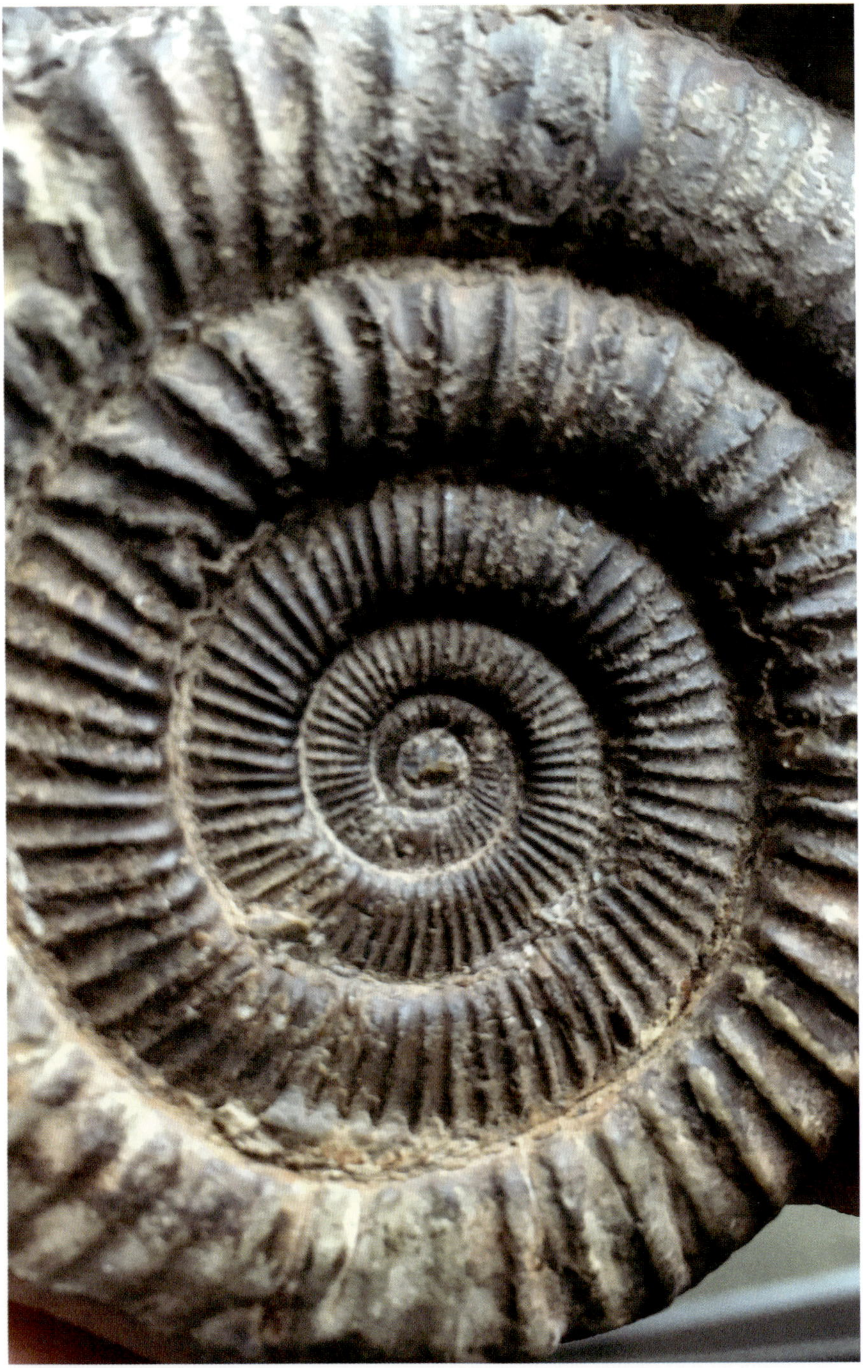

RUNSWICK BAY, NORTH YORKSHIRE

SSSI Site – no hammering or digging in the cliffs or bedrock

The Early to Late Liassic rocks at Runswick Bay, coupled with ease of access, foreshore collecting and a high frequency of finds, make this a very good location for families trying their hand at fossil collection. The early Jurassic here is represented by Pliensbachian age rocks (195–190 mya) from the Early Lias and Toarcian age (190–180 mya) from the Late Lias.

This Lower Jurassic site is in the Upper Lias of the Whitby Mudstone Formation, consisting of deposits of the Grey Shale Member, the Mulgrave Shale Member and the Alum Shale Member. The Lower Lias consists of the Cleveland Ironstone Formation, whose deposits consist of the Penny Nab Member and the Kettleness Member.

UPPER LIAS	TOARCIAN	WHITBY MUDSTONE FORMATION

- ALUM SHALE MEMBER
- MULGRAVE SHALE MEMBER
- GREY SHALE MEMBER

LOWER LIAS	PLIENSBACHIAN	CLEVELAND IRONSTONE FORMATION

- KETTLENESS MEMBER
- PENNY NAB MEMBER

Undoubtedly, the coast of North Yorkshire and surrounding the town of Whitby, is a mecca for Jurassic marine fossil collectors. Although any section of coast from Staithes to Ravenscar will offer much the same fauna, especially well preserved ammonites at Runswick Bay. This site is about 9 miles from Whitby and is very suitable for families.

Access to the beach is easy, with car parking close by. Head east, once on the beach search among the rocks and beach rubble that occurs all the way to Kettleness. Take care to watch the tide, as there is a danger of being cut off!

The secret at this location is to find the hard, ovate or spherical shaped nodules, which contain very well preserved ammonites. These nodules erode because of the sea's action upon them and the keel of the ammonite within can often be seen. Sometimes, the nodules can be found in situ protruding from fallen blocks of shale on the beach and can simply be collected from the bedrock.

The nodules need to be 'cracked' open, using a geological hammer. This is a relatively easy task and a well-aimed blow should provide a good specimen and the negative impression too. Ammonites are the most common fossil found in the nodules here but other fossils can also be found at Runswick Bay, including the large oyster *Gryphaea*, various molluscs and belemnites. Remains of Pleistocene mammals, such as mammoth or bison from the North Sea are also often found here, washed up on rough tides.

South of Runswick Bay the beach gives way to cliffs and a rocky foreshore.

The east end of the beach comprises rocks and rubble.

Ammonites, such as *Dactylioceras*, are the most common fossils. Ensure you wear goggles when attempting to break the nodules.

Typical nodule found at Runswick Bay, in this case containing a good-sized ammonite.

Dactylioceras tenuicostatum found by Tracey Jones on a UKAFH hunt.

Site summary

⚠️ Rock falls

⚠️ Slippery rocks

★★★★★ Accessibility

★★★★ Child suitability

★★★★ Find frequency

SSSI site. No hammering or digging in the cliffs or bedrock.

Easy parking. Crumbling, unstable cliffs. Do not attempt to climb. Rocky foreshore can be slippery. Be aware of tides.

Suggested equipment: Geological hammer, chisels, goggles, specimen box, wrapping material.

Directions: Follow the A174 south from Hinderwell, turning left to Runswick Bay Hotel and the bay.

Postcode: TS13 5HT

UKAFH hunt at Runswick Bay in February 2016.

LAVERNOCK, SOUTH WALES

SSSI Site – no hammering or digging in the cliffs or bedrock

Lavernock Point is located a little to the west of Cardiff, at the edge of the mouth of the River Severn. It is a small place that attracts an inordinate number of fossil hunters, due to its more-than-generous supply of fossils. It is formed from layers of the Lower Blue Lias limestone and shales. The Jurassic rocks here are similar to those found at Watchet.

Fossils to be found here include bones or teeth from marine reptiles, particularly ichthyosaurs and plesiosaurs. Ammonites, large gastropods, bivalves and brachiopods can also be collected. The Triassic rocks here are unfossiliferous and belong to the Mercia Mudstone Group (formerly named the Sully Beds, now the Blue Anchor Formation).

Cliff section at Lavernock Point.

Stepped beds on the foreshore at Lavernock.

Bones of the dinosaur *Dracoraptor hanigani*.

When approaching Lavernock Point from the direction of Penarth, the Blue Anchor Formation forms a prominent cliff beyond which are some new sea defences made from massive blocks of hard, grey Carboniferous Limestone, brought in from outside the area.

Between here and the Point, the Westbury Formation is well exposed in the cliff and foreshore. It contains a thin irregular layer of sandstone rich in the mineral iron pyrites and pieces of a bone bed, containing bones and teeth of fish and marine reptiles. Slabs of these can usually be found on the beach.

Westwards from Lavernock Point the Blue Lias is exposed and the cliffs of St. Mary's Well Bay are folded in a gentle syncline; its three divisions (St. Mary's Well Bay Formation, Lavernock Shales and Porthkerry Formation) can easily be recognised.

At the Point the bedding planes of the limestones contain large groups of the oyster *Liostrea*. Walking westwards, and therefore onto younger beds, the flattened, coiled shells of the ammonite *Psiloceras* are quite common.

In June 2015 it was announced that the fossil remans of a new species of dinosaur, since named *Dracoraptor hanigani*, had been discovered at Lavernock Point. (see pp.136–137) and is believed to be the earliest known Jurassic theropod dinosaur.

Model of *D. hanigani* by Bob Nichols, of paleocreations.com.

Dracoraptor hanigani – a Welsh dinosaur from Lavernock

The Blue Lias Formation at Lavernock Point is a constant source of fossils. The Point is subjected to rapid erosion from both sea and weather, causing the cliffs to become incredibly unstable. Rock falls are commonplace and local fossil collectors regularly search the foreshore and at the base of the cliffs, for the quality specimens that occur here. One such collector is Nick Hanigan, who along with his brother Rob, has been collecting from the fallen blocks of Late Triassic and Early Jurassic marine strata at Lavernock Point since childhood.

In March 2015, a partial disarticulated skeleton, including the skull, of a new genus and species of theropod dinosaur was discovered within a minor cliff collapse of well bedded limestones and mudstones at Lavernock Point. Its occurrence close to the base of the Blue Lias Formation (Lower Jurassic, Hettangian) makes it the oldest known Jurassic dinosaur and it represents the first dinosaur skeleton from the Jurassic of Wales. It was a distant cousin of *Tyrannosaurus rex* and lived 201.3 ± 0.2 million years ago. It is also one of the most complete theropods from the European Lower Jurassic.

The name, *Dracoraptor hanigani*, credits the two brothers for their important discovery. *Dracoraptor* was a small, agile carnivore, about 2.3 feet (70 cm) tall and 6.5 feet (2 m) long, with a long tail. The fossil is thought to be from a juvenile animal, as most of its bones were not fully formed.

The bones of the theropod dinosaur, *Dracorapor hanigani*, prepared by Craig Chivers, who kindly supplied the photos.

Reconstruction of *Dracoraptor* and (below) more of the skeleton and a tooth of the first Welsh dinosaur.

Associated fauna of echinoderms and bivalves indicate that the specimen had drifted out to sea, presumably from the nearby Welsh Massif and associated islands (St. David's Archipelago). Theropod dinosaurs are extremely rare in the Lower Jurassic and most reports are of only fragmentary remains. This rarity results in a considerable gap in our knowledge of these animals at a time when, indications are, theropods were diversifying rapidly.

Approximately 40% of the animal was discovered by the Hanigan brothers. The larger blocks collected were dried over two weeks under damp newspaper to prevent rapid shrinkage and cracking of a thin mudstone veneer on the limestone surface.

In 2015, Sam Davies, a University of Portsmouth palaeontology student, discovered the animal's fossilised foot during a field trip to the very same spot. The skeleton and reconstruction of *Dracoraptor haniganii* now reside in a display at the National Museum Wales, in Cardiff.

The authors gratefully acknowledge the contributions made to this section of the book by Nick Hanigan, Sam Davies, Craig Chivers and Bob Nichols.

Brittle star from Lavenock beach.

Large plesiosaur vertebra from Lavernock.

Site summary

⚠️ Rock falls

⚠️ Slippery rocks

★★★ Accessibility

★★★ Child suitability

★★★★ Find frequency

SSSI site. No hammering or digging in the cliffs or bedrock.

Best for older children due to crumbling, unstable cliffs. Good access to the beach and easy parking for up to four cars.

Suggested equipment: Geological hammer, chisels, goggles, specimen box, wrapping material.

Directions: From the B2467, follow signs to Lavernock and then to Fort Road (a no-through road). At the end of the lane is a church and parking for up to four cars. A narrow path leads to the beach.

Nearest postcode: CF64 5UL

WATCHET, SOMERSET

SSSI Site – no hammering or digging in the cliffs or bedrock

The Hettangian Formation of the Early Blue Lias is exposed along the coast east of Watchet. The cliffs to the west belong to the Mercia Mudstone Group of the Late Triassic period and are relatively unfossiliferous. The Lias, however, is rich in fossils but usually only after scouring conditions.

Ammonites and marine reptile remains can be found among the sequences of shale and limestone and the rock platforms, which extend across most of the foreshore. These can provide good fossil hunting grounds given the right conditions, such as after strong tides during the winter months. Collecting good fossil specimens at Watchet is often a matter of luck, as this popular site can be over-collected and is visited daily by the locals.

- EARLY JURASSIC — HETTANGIAN — BLUE LIAS FORMATION
- LATE TRIASSIC — PENARTH GROUP — WESTBURY FORMATION
- MIDDLE TRIASSIC — MERCIA MUDSTONE GROUP — BLUE ANCHOR FORMATION

This section of the Somerset coast is of great interest to geologists, as there are excellent examples of faulting and unconformities. The cliffs in some parts of this coastline are under constant attack from the sea and erode readily but at different rates, causing interesting geological features.

Under normal conditions, there is very little to be found at Watchet these days and it is true to say that a good storm, with scouring tides or a new cliff fall are certainly needed to find the better specimens. Collecting can then be productive but the site is overworked by collectors. It is invariably picked clean of fossils on an almost daily basis, with even smaller and inconsequential specimens getting picked up! Watchet is certainly not as prolific in fossils, as some books and websites often suggest.

The dramatic cliffs at Watchet extend along the North Somerset close to Kilve and Quantoxhead.

The strata here dips seaward and where the sea platforms slope up and face the inland side, they can be very efficient at collecting nodules, cobbles and shingle. When these are scoured in the spring or autumn tides, they can sometimes reveal reptile bones.

Ammonites are relatively common in the shales, represented by *Psiloceras planorbis* and there are also trace fossils to be seen but these are almost impossible to extract and are best left alone or photographed. However, ammonites can be recovered with care but these, like reptile bones, are best preserved when they are found in nodules.

The foreshore ledges are the best place to look for fossils.

The ammonite, *Psiloceras planorbis*.

Typical ammonite finds among the ledges at Watchet.

Site summary

⚠️ Rock falls

⚠️ Slippery rocks

★★★★ Accessibility
★★★★ Child suitability
★★ Find frequency

SSSI site. No hammering or digging in the cliffs or bedrock.

Rocky foreshore. Frequent falls from cliffs. Avoid the base of cliffs.

Suggested equipment: Geological hammer, chisels, goggles, wrapping material.

Directions: Access to the beach at Watchet is mostly by parking in the town itself, at one of nine available car parks and walking from Watchet Harbour along the beach. The car park at Harbour Road is situated at postcode: TA23 0AQ

Nodules once were a common occurrence on the foreshore at Watchet, trapped in the ledges having fallen from the cliffs. The trick these days is to actually find one! Approach a visit to Watchet with a view to seeing some spectacular coastal scenery and to finding a few fossils as a bonus!

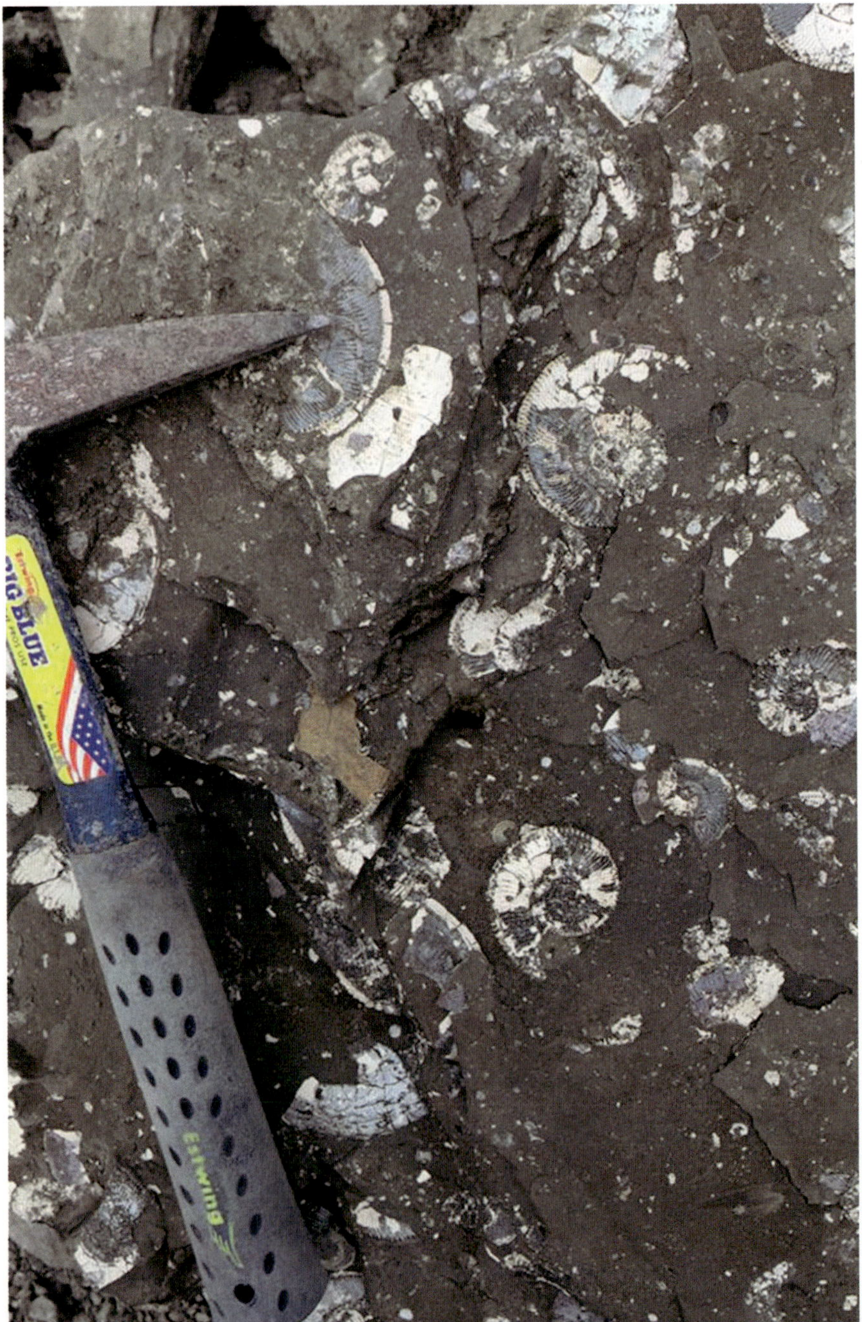
Estwing

KING'S DYKE NATURE RESERVE, WHITTLESEY, CAMBRIDGESHIRE

This site is set within the King's Dyke Nature Reserve, in a special fossil hunting area provided by the adjacent brickworks, located next to it. The main quarry is situated in the highly fossiliferous Peterborough Member Formation of the Lower Oxford Clay (Callovian age of 166.1–163.5 mya), where the clay is extracted for brick manufacture and the brick pit regularly provides the fossil area with fresh material from it's spoil tip.

The main quarry has been the source of some excellent specimens of marine reptiles, fish and more over the years but access to the pit is no longer possible for the general public. The fossil hunting area, within a disused quarry is especially ideal for families, with older children and will provide a wealth of fossils for all.

MIDDLE JURASSIC | CALLOVIAN | LOWER OXFORD CLAY | PETERBOROUGH MEMBER FORMATION

Humerus of a plesiosaur, possibly *Cryptocidus*. Bones are a rare find in the fossil hunting area but are frequently found in the main quarry.

***Keplerites* ammonite found at King's Dyke Nature Reserve.**

Belemnites *Cylindroteuthis puzosiana*.

It needs to be emphasised that access to the reserve is through a free permit only. There is no access to any other part of the quarry, only the fossil hunting area.

The area provides safe fossil collecting and is ideal for families with older children. Ammonites (especially *Kosmoceras jason* and *Keplerites* sp.), belemnites (*Cylindroteuthis puzosiana*), fossilised wood and the oyster *Gryphaea dilobotes*, are all very common finds. They occur on the surface of the spoil tip with good frequency.

The thin, compacted layers of the Oxford Clay at this site can be split easily using a bolster chisel or wallpaper scraper. Ammonites are very fragile and will need to be wrapped in tissue and bubble wrap, then varnished at home to prevent them from drying out and disintegrating. PVA diluted with water 1:3 is also a good overall preservative. This can be applied on site if necessary, if a solution is brought along in a plastic bottle and the specimen is allowed to dry before wrapping.

Worm burrows, gastropods and crinoids are also found here, along with shark teeth, and marine reptile teeth and bones (these include plesiosaurs, ichthyosaurs, pliosaurs and crocodiles). As with all spoil heap collecting, the fossils are not found within their in situ horizons, which means if you find a significant piece of bone, for example, it will be highly unlikely to find any more that might complete the specimen!

Kosmoceras ammonites from King's Dyke are invariably crushed and friable.

Kosmoceras jason.

The oyster *Gryphaea dilobotes* is a common find.

Site summary

★★★★★ Accessibility

★★★★★ Child suitability

★★★★☆ Find frequency

Suitable for families with older children. Easy car parking with nearby toilet. No restrictions.

Suggested equipment: Geological hammer, flat splitting chisel or wallpaper scraper (for splitting the clay), steel pick, wrapping material.

Directions: The reserve is situated off the A605 between Peterborough and Whittlesey. Address is Kings Dyke Nature Reserve, 222 Peterborough Road, Whittlesey, Peterborough, Cambridgeshire.

Postcode: PE7 1PD

Access is free but you must ensure that you obtain a permit, in advance, from philipparkerassociates@btconnect.com or through the King's Dyke Nature Reserve website at www.kingsdykenaturereserve.com.

HAMPTON VALE LAKE, YAXLEY, CAMBRIDGESHIRE

SSSI Site – no hammering or digging in the banks

The lakes around the village of Yaxley are the result of the extensive brick working and clay quarrying that took place in the area. Much of this industry is now confined to a few productive pits and in the case of Hampton Vale, the land now comprises a new housing estate, with the benefit of a large country park, in which Hampton Vale Lake is set.

On the southeastern banks of the lake, the Lower Oxford Clay of the Peterborough Member Formation (Callovian age of 165–159 mya) is exposed. Access to the banks is straightforward, from the public footpath that circumnavigates the lake and the base of the embankments can be very productive for fossil collection.

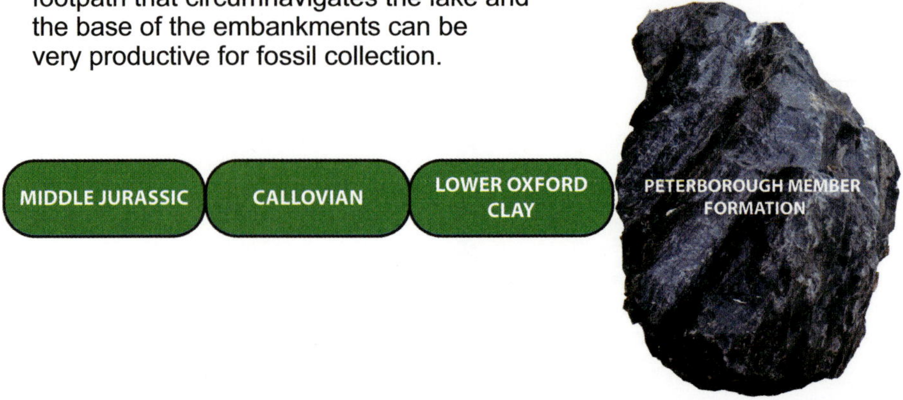

- MIDDLE JURASSIC
- CALLOVIAN
- LOWER OXFORD CLAY
- PETERBOROUGH MEMBER FORMATION

Ichthyosaur vertebra.

Fossil hunting on the banks of the lake.

Pyritised ammonites from the banks of Hampton Lake.

This location is probably not suitable for young children, as it is near to deep water. Parental supervision for any non-adult is strongly advised. The banks of the lake are constantly eroding and collecting is best at the base of the embankment during summer months, when the water levels have significantly dropped.

Fossils to be found here are similar to those found at the King's Dyke Nature Reserve site (see previous section) but the remains of marine reptiles are surprisingly common. Expect to find ammonites, belemnites, crinoid stems, the oyster *Gryphaea dilobotes*, worm casts, gastropods and other bivalves.

Many of the ammonites here are pyritised, which are different from those found at King's Dyke. These ammonites need to be cleaned and are best preserved by a coating of varnish (water soluble artist's varnish is ideal) and kept in a drawer or container with a sachet of silica gel, to help prevent further pyrite decay.

Pyrite decay is not easily prevented. Specimens that consist almost wholly of pyrite are easily identified as such, as they look and feel 'metallic'. The oxidation of the pyrite (FeS_2) causes the problem and fossils, over time, may disintegrate.

One useful means of finding fossils in the clay at Yaxley, is to use wet sieving. This is particularly easy, with plenty of water at hand and can reveal a good number of smaller ammonites, which are easily overlooked. A quarter or a half inch mesh is ideal.

Tooth of *Simolestes*, a pliosaur

Hampton Vale Lake, showing the Oxford Clay embankments and typical ammonites found here.

Site summary

Deep water

★★★★★ Accessibility

★★ Child suitability

★★★★ Find frequency

SSSI site. No hammering or digging in the banks.

Location is close to deep water therefore not suitable for younger children. Child supervision is advised. Car parking in nearby residential estate.

Suggested equipment: Small steel pick, sieve, trowel, specimen box, wrapping material.

Directions: From the A1139 Peterborough bypass, take the A1260 towards Yaxley. Turn off to Hempton Villa, the new housing estate and park along the road within the estate.

Postcode: PE7 1PD

Access is free.

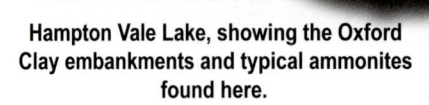

BLACK VEN, CHARMOUTH, DORSET

SSSI Site – no hammering or digging in the cliffs

This is quite possibly one of the most famous fossil sites in the world and with good reason. It was to here that Mary Anning walked from Lyme Regis, exploring the constantly crumbling cliffs along a stretch of the Jurassic Coast, now famed for its magnificent fossils.

Charmouth's beach, even on a wet day not fit for walking the dog, is usually straddled with people, doubled over looking for fossils! Black Ven lies to the west of Charmouth and the cliffs are constantly moving; ravaged and eroded by tides and in a highly unstable condition all year. Mudslides, landslips and cliff falls are a familiar sight, so keep clear of the cliff base. At Charmouth's Black Ven, the Lias of Dorset is exposed and the profile of the cliff is as follows:

- LOWER CRETACEOUS
  - LATE ALBIAN
    - UPPER GREENSAND
    - GAULT CLAY
  - MIDDLE ALBIAN
    - GREEN AMMONITE BED
    - BELEMNITE MARLS
- JURASSIC
  - MIDDLE LIAS
    - BLACK VEN MARLS
      Birchi nodules at base
    - SHALES-WITH-BEEF
    - BLUE LIAS

So, where to start? Luckily, the car park at Charmouth is very close to the beach and nearby are toilets, a cafe and the Charmouth Fossil Heritage Information Centre. The Heritage Centre is a good place to start, as it enables collectors a good opportunity to see the types of fossils that can be found here. From the Heritage Centre, continue westwards under Black Ven towards Lyme Regis.

At Black Ven, almost immediately you will see the beds called Shales-with-Beef, at the base of the cliff, which is named from the abundance (at some levels) of 'beef' (a vertically oriented calcite). At the top of the succession are two particularly obvious beds of limestone, which are the spherical concretions of the Birchi Nodular, overlain by the Birchi Tabular, which can yield fine ammonites, like the well preserved and uncompressed specimens of the ammonite *Microderoceras birchi*, much sought by local collectors. Fossil insects can also be found here. The Black Ven Marls follow.

Image previous page: © Nick Macneill. Image top: © Nigel Mykura. Image bottom: ©N Chadwick. Works are licensed under the Creative Commons Attribution.

The high, land-slipped and often boggy cliff, at Black Ven, has mudslides (mudflows) which often project into the sea at the base. The beds in the cliffs above contain many fossils, including ammonites and specimens can be brought down in the landslides, to be washed out on the beach.

We strongly advise the collection of fossils only from the beach and foreshore and not to attempt to either climb the cliffs or extract fossils from them. The cliffs are highly dangerous and prone to collapse at all times. The site is SSSI designated and should not be hammered or dug into under any circumstances.

The Black Venn Marls, seen above the Birchi Tabular are part of the Charmouth Mudstone Formation of the Lias Group. The Marls contain harder bands, which may contain uncompressed ammonites. Above, are paper shales with the ammonite *Asteroceras obtusum* and then

Androgynoceras from the Green Ammonite Beds at Black Ven. Photo courtesy of Allhallows Museum, Honiton.

Flatstones and Woodstone. The Flatstones contain an interesting fauna of insects, including grasshoppers, beetles and dragonflies and the dinosaur, *Scelidosaurus* was found in this layer, a cast of which can be seen in the Charmouth Heritage Centre.

Then there is an impersistent Pentacrinite Bed, with the magnificently preserved crinoid *Pentacrinus fossilis*, followed by an impersistent limestone with the ammonite *Promicroceras planicosta*. The stellare Nodules are septarian and can contain the common ammonite *Asteroceras stellare* in an uncrushed condition.

The clearly observed second terrace on Black Ven is formed by the clays above the stellare Nodules. The lymense Bed, a pyritic horizon with beef and ammonites occurs above the stellare Nodules and further up is Watch Ammonite Stone, a limestone with ammonites of *Echioceras raricostatum*, and finally at the top of the Black Ven Marls there is a thin pyritic limestone known as Hummocky.

The majority of the beds on Black Venn are inaccessible. The crumbling sheer cliffs, fraught with danger of collapse and mudslide makes fossil collecting restricted to the beach, where ample specimens can be found under the right conditions and with perseverance.

Asteroceras obtusum.

Asteroceras stellare from a stellare septarian nodule.

An organised UKAFH fossil group with Dean Lomax on the foreshore at Black Ven.

152

Pentacrinites crinoid block found at Black Ven, (along with an ichthyosaur rostrum) by Brandon Lennon. Image credits: Brandon Lennon. www.lymeregisfossilwalks.com

A rare find! A dragonfly wing *Liassophlebia* sp. from Black Ven belonging to David Penney.

Site summary

⚠ Rock falls

⚠ Slippery rocks

★★★★★ Accessibility

★★ Child suitability

★★★ Find frequency

SSSI Site. No hammering or digging into the cliffs of this protected Jurassic Coast World Heritage Site.

The dangerous crumbling clay cliffs and rocky foreshore make this a site suitable for older children only. Foreshore collection strongly advised. Stay away from cliff base and do not attempt to climb the cliffs.

Suggested equipment: Geological hammer and chisels, specimen box, wrapping material.

Directions: Head into Charmouth and turn into Lower Sea Lane. Adequate parking can be found in car parks close to the Charmouth Heritage Centre. Head west to Black Ven.

Postcode: DT6 6LL

The fast eroding and crumbling cliffs below Stonebarrow Hill arefamed for their ammonite fossils but marine reptile remains are also common, especially the vertebrae of ichthyosaurs. The foreshore is the safest place for collection.

EAST BEACH & STONEBARROW, CHARMOUTH, DORSET

SSSI Site – no hammering or digging into the cliffs

East Beach and Stonebarrow at Charmouth are a continuation of the strata found west of the River Char at Black Ven. Here, faulting has occurred, resulting in the rock layers present higher in the cliffs at Black Ven, being brought down to beach level. Fossils can be found along the entire stretch of beach, especially within the first 1 km of the beach access point and at low tide in the shingle and exposed foreshore. Thousands of visitors scour the beach annually for fossils, especially pyrite ammonites, washed out of these famous cliffs.

Charmouth's East Beach, below Stonebarrow Hill is a popular spot and the fossils can be simply picked from the beach, although winter hunting avoids the crowds and find frequency! On Charmouth's East Beach below Stonebarrow Hill, the Lias of Dorset is exposed and the profile of the cliff is as follows:

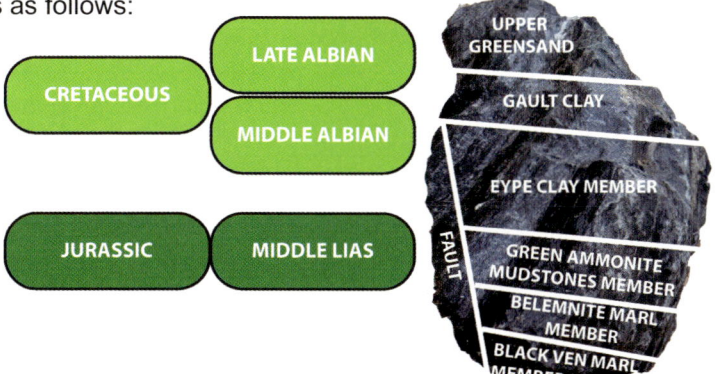

We strongly advise the collection of fossils only from the beach and foreshore and not to attempt to either climb the cliffs or extract fossils from them. The cliffs are highly dangerous and prone to collapse at all times. The site is an SSSI and should not be hammered or dug into under any circumstances.

The photo on the top left shows a 2016 landslip that brought down several tons of clay and rock and illustrates how unpredictable and hazardous these cliffs are. Mudflows are also common at both East Beach and Black Ven, which are also very dangerous.

Landslip on East Beach, February 2016.

Promicroceras planicosta cluster.

East Beach is a great site for children, particularly if you carefully adhere to the safety conditions that this site warrants. Staying on the beach, away from the cliffs and looking along the foreshore can be rewarding, especially during the winter months and where the sea has already worked on fresh material that has fallen from the cliffs. Pyritised ammonites are the 'prize' for most! However, reptile remains, crinoid and fish remains also occur with good frequency.

East Beach is a popular site. There is every chance that you might find nothing, as the beach is so populated with fossil collectors and it can get picked clean almost daily. An early start, following out a low tide will certainly help.

Cliff collapses at East Beach are frequent. Avoid the base of the cliffs and let these do all the hard work for you.

Beneath Stonebarrow, the lowest and oldest sediment is that of the Black Ven Marl Member, comprising of mostly dark-grey mudstones, with subordinate beds of nodular and tabular limestone. A 0.3 metre-thick limestone bed, the Lower Cement Bed, forms a conspicuous marker within the Black Ven Marls and is visible in the lower part of the cliff. This disappears beneath the shingle about 1,500 metres from the beach access point.

Overlying the Lower Cement Bed is the Upper Cement Bed, identifiable by two closely spaced limestone bands. These dark-coloured sediments are largely responsible for the volume of pyrite ammonites that occur on the foreshore. These include: *Crucilobiceras*, *Eoderoceras*, *Echioceras* and *Oxynoticeras lymense*.

Towards the top of the cliff can be seen the Belemnite Marl Member. Continuing along the beach (towards Seatown) you eventually reach the cliffs beneath Golden Cap, where the Green Ammonite Mudstone Member is exposed in the lower part of the cliff, within which very well preserved ammonites can be collected.

Echioceras ammonites from the fall in February 2016.

Site summary

Rock falls

Slippery rocks

★★★★★ Accessibility

★★★★ Child suitability

★★★★ Find frequency

SSSI Site. No hammering or digging into the cliffs of this protected Jurassic Coast World Heritage Site.

The dangerous crumbling clay cliffs and mudflows are hazardous. Stay well away from cliff base. Foreshore beach collection is safe for children.

Suggested equipment: Geological hammer and chisels, specimen box, wrapping material.

Directions: Head into Charmouth and turn into Lower Sea Lane. Adequate parking can be found in car parks close to the Charmouth Heritage Centre. Turn east for East Beach and Stonebarrow.

Postcode: DT6 6LL

158

SALTWICK BAY, NORTH YORKSHIRE

SSSI Site – no hammering or digging into the cliffs or bedrock

Despite the rich rewards of fossils found on the foreshore, Saltwick Bay is not a site suitable for young children. Situated just east of Whitby, accessibility can be difficult and the steps leading to the beach are hazardous in stormy weather. Fossils can be found in nodules on the beach, or loose within the shingle and shale. Ammonites are plentiful here but the site can be over-collected and the marine reptile remains are not particularly common as a result.

The Alum Shale Member contains molluscs and ammonites, particularly *Dactylioceras* and *Hildoceras*, whilst the overlying Dogger Group includes the Whitby Plant Beds, within the fine-grained sandstone blocks.

The access stairs are located just past the holiday camp facilities which include a small store and a snack bar. The descent to the beach is a bit arduous!

Like so many coastal locations, fossil collecting at Saltwick Bay is best conducted along the foreshore. Look between rocks and boulders for ammonite nodules, which can be broken with a geological hammer, to reveal beautiful ammonite specimens. The nodules are common enough and fall from the Alum Shale Member above. The bivalves *Dacryomya ovum* and *Pleuromya* sp. are a common find, along with belemnites. The long thin ones (*Cuspiteuthis tubularis*) can be found on the surface of the shale near to Black Nab at Saltwick Bay.

Plant remains are quite common in the Whitby and Scarborough areas and Saltwick Bay is no exception. The Dogger Group of fine-grained sandstones reaches down to beach level here and the Whitby Plant bed reveals leaves, usually preserved as carbon impressions of the original leaf. Tree trunks and branches are sometimes preserved as flattened coal like bands, sometimes as mineralised casts and occasionally as Jet.

Collecting fossils at Saltwick Bay can be very dangerous. The cliffs are constantly losing pieces, ranging from small flakes of shale to massive boulders of sandstone.

UKAFH fossil hunt at Saltwick Bay, with Dean Lomax.

Nodules containing ammonites.
Image credit: Ted Gray.

Jurassic lobster, *Pseudoglyphea* sp.
Image credit: Ted Gray.

Pyritised belemnite and phragmacone,
collected by Ted Gray.
Image credit: Ted Gray.

Site summary

⚠ Rock falls

⚠ Slippery rocks

★★★★★ Accessibility
★★★ Child suitability
★★★★ Find frequency

SSSI Site. No hammering or digging into the cliffs or bedrock.

Dangerous cliffs.
Tidal location.
Steep descent to beach level.

ALWAYS wear protective goggles and particularly if you try to break open an ironstone nodule.

Suggested equipment: Geological hammer, chisels, goggles, wrapping materials, rucksack.

Directions: Follow the coastal road from Saltwick Bay Abbey and park outside the Whitby Holiday Park. Walk to the steps to the foreshore.

Postcode: YO22 4JX

PENARTH, SOUTH WALES

SSSI Site – no hammering or digging into the cliffs or bedrock

The Blue Lias rocks at Penarth are similar to those found at Sedbury Cliff, Quantoxhead, Watchet and Lavernock. This is hardly surprising, as the Jurassic rocks at Penarth, extend across the Bristol Channel but differ in terms of the fauna found. At Penarth, brachiopods, bivalves and gastropods are found in higher numbers, with fewer ammonites present.

The Triassic red cliffs here are of Ladinian age (233–227 mya) and belong to the Mercia Mudstone Group, with Rhaetian (219–206 mya) mudstones, limestones and sandstones of the Penarth Group above. Both are largely unfossiliferous. The Jurassic Lower Lias of Hettangian age (205.7–201 mya) provides collecting from the Lavernock Shale Member, found as boulders and shale on the foreshore.

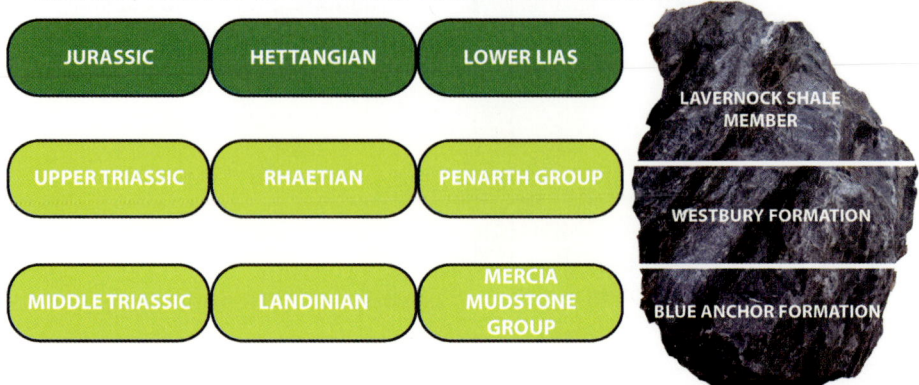

JURASSIC	HETTANGIAN	LOWER LIAS
UPPER TRIASSIC	RHAETIAN	PENARTH GROUP
MIDDLE TRIASSIC	LANDINIAN	MERCIA MUDSTONE GROUP

LAVERNOCK SHALE MEMBER

WESTBURY FORMATION

BLUE ANCHOR FORMATION

Penarth is a very good collecting locality and is a highly popular site. It can be busy here and the site can be over-collected. However, the high rate of erosion keeps the foreshore topped up and most collectors will be able to obtain a good collection from the section between Penarth Pier and Cardiff Barrage.

With easy access to the beach, Penarth provides a good site for the family but stay clear of the cliffs and cliff base. The cliffs are unstable, so collection is always best on the foreshore, carefully looking in the fallen blocks and where necessary, using a geological hammer to break the rock. Always use goggles when doing so.

Reptile remains are not uncommon here and vertebrae of ichthyosaurs and plesiosaurs often turn up. In 2015, a 7ft.-long (2.1 metres) ichthyosaur skeleton was uncovered on Penarth beach.

For most people, it is suggested that you start from Penarth Pier, walking along the beach towards Cardiff Barrage. Fossils can be found along this entire stretch of coast and the chances of finding bivalves, gastropods or brachiopods is high.

Please be aware that this is a tidal location and the tide can come in fast.

Views of Penarth beach.

Vertebrae and ribs of an ichthyosaur.

Brittle stars.

A large, well preserved lobster.

Site summary

⚠️ Rock falls

⚠️ Slippery rocks

★★★★★ Accessibility

★★★★★ Child suitability

★★★★ Find frequency

SSSI Site. No hammering or digging into the cliffs or bedrock.

Suggested equipment: Geological hammer, chisels, goggles, wrapping materials, rucksack.

Directions: Parking along Penarth seafront (with cafe and toilets) during winter months.

Postcodes: Cliff Parade car park CF64 5YY, Penarth Portway car park CF64 1TS

IRCHESTER COUNTRY PARK, NORTHAMPTONSHIRE

This former ironstone and sandstone quarry now forms Irchester Country Park. The quarrying has exposed the Jurassic geology and the site is now given RIGS (Regionally Important Geology Site) status.

The iron ore is referred to as 'Northampton Sands with Ironstone'. It was up to 7 metres thick in places and underlies a succession of sands, overlain by mudrocks, clays and limestones, amounting to a depth of about 16 metres. These strata are all Jurassic in age, being laid down between 177–170 mya. The sand was deposited under the sea during the Bajocian stage of the Middle Jurassic.

Blocks of the fossiliferous limestone rocks can be found lying in the quarry and regular falls ensure a supply of fresh material.

- JURASSIC
- BAJOCIAN
- INFERIOR OOLITE

NORTHAMPTON SAND FORMATION (FORMERLY NORTHAMPTON SAND IRONSTONE FORMATION)

Jurassic-aged old quarries are strewn across Northamptonshire and Irchester Country Park is one of them. Easily accessible, it provides a good fossil site but is not suitable for children. The problem of potential rockfalls or occasional flooding and overgrown vegetation are very real! We strongly recommend the wearing of hard hats at this site. The old quarry face is accessible by way of footpaths through the park, which are well signposted.

Fossils are found in fallen blocks and may contain ammonites, wood, bivalves (*Modiolus*, *Pholadomya*, *Lopha*, *Gryphaea*, etc.), ostracods, calcareous worm tubes, belemnites, gastropods (both nerineid and others), brachiopods, bryozoa, echinoids (*Clypeus* and others), crinoids and corals. Trace fossils include *Trypanites*, *Gastrochaenolites*, *Thalassinoides* and other burrows. Brachiopods and bivalves can also be found loose on the ground.

The old quarry face at Irchester Country Park, which is prone to overgrown vegetation during the summer months.

Groups and individuals, as a courtesy to the park, need to book in with the Park Rangers prior to their visit. Telephone 0300 12 65934. This will help with monitoring the numbers using the site, updating the rangers on where finds are being made and help provide useful contacts for future park events.

Base of Northampton Sand ironstone with various bivalves and trace fossils of *Thalassinoides*.

The rangers in turn can help users to understand the basic rules of the park regarding fossil hunting, provide updates on the condition of the quarry face and avoid clashes with other groups or members of the public.

Site summary

⚠️ Rock falls

⚠️ Slippery rocks

★★★ Accessibility

★ Child suitability

★★★★ Find frequency

Hard hats should be worn, as rock falls frequently occur.

Walking boots recommended. Site not suitable for children.

Directions: Irchester Country Park, Gipsy Lane, Little Irchester, Wellingborough, Northants.

Book in with the Park Rangers prior to visit. Telephone 0300 12 65934.

Postcode: NN29 7DL

A selection of fossil bivalves and brachiopods from Irchester Country Park.

167

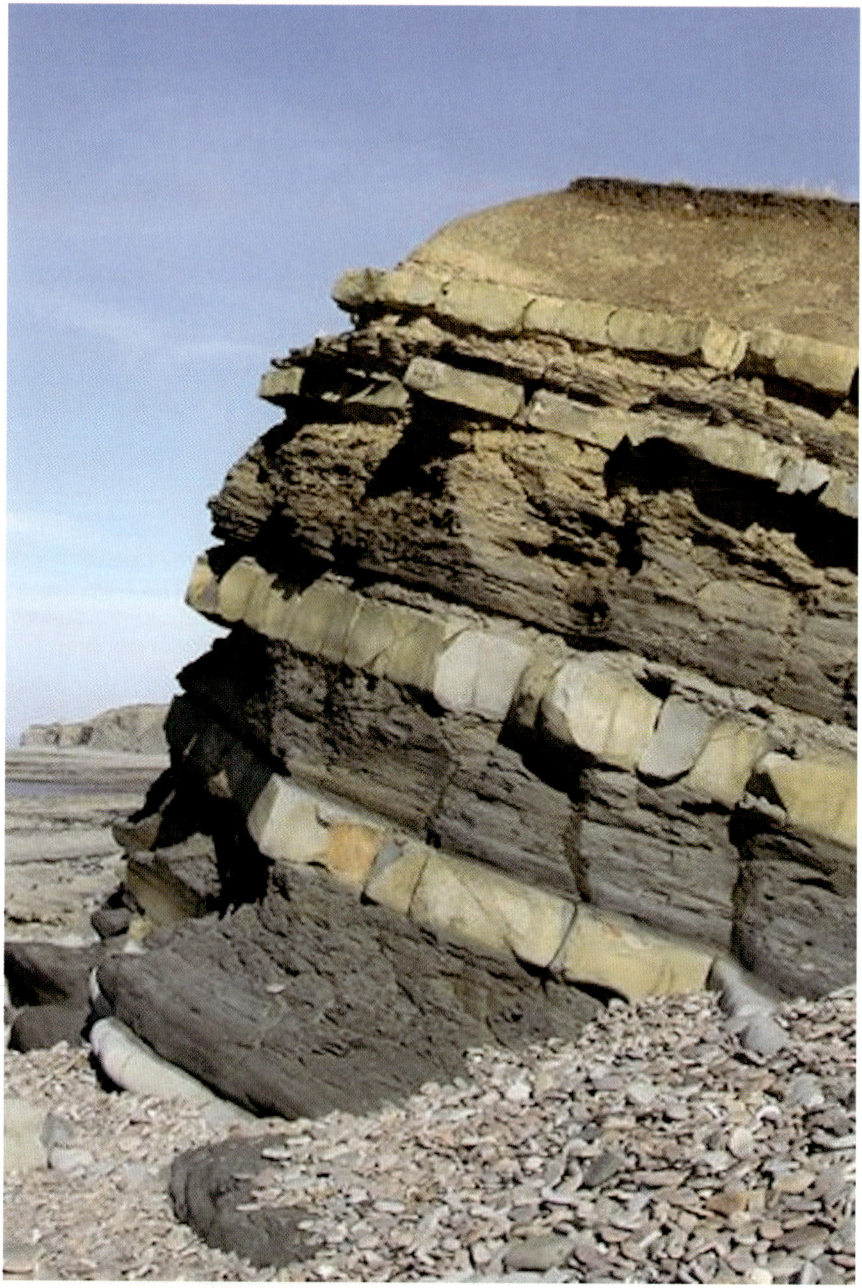

QUANTOXHEAD, SOMERSET

SSSI Site – no hammering or digging into the cliffs

Dipping eastwards, the Lower Lias clays and limestone at Quantoxhead, situated on the Somerset coast, are similar to those found at Church Cliff at Lyme Regis. The miles of tall Jurassic cliffs and the spectacular wave-cut platform can provide a rich source of fossils, if conditions are right.

Ammonites can be found in nodules and loose within the hard limestone layers or loose on the beach. Impressions are also common along this section. However, fossils do not necessarily abound and even with the Lias, the shore never really lives up to its reputation of being rich in either a shelley fauna or reptile remains, especially. It is a popular location and can be frequently picked clean by others.

- JURASSIC
  - SINEMURIAN
  - HETTANGIAN
- EARLY LIAS
- BLUE LIAS FORMATION

The coastline from Watchet, eastwards through Kilve and onto Quantoxhead is composed of the same Lower Lias clays and limestone, giving a continuous and impressive section of layered cliffs. As with Watchet, fossils are found mostly on the wave cut platform that extends along the shoreline and the frequency of finds seems to increase west of the steps down to the beach. Here, ammonite impressions are common but you will need a fresh cliff fall to find whole ammonites or the nodules in which they are often preserved.

Marine reptiles remains are not as frequent as is often professed. This is a well-attended stretch of coast by collectors and is visited daily; it can be disappointingly picked clean of the better finds.

The impressive Lower Lias clays and limestone cliffs and foreshore at Quantoxhead.

Other fossils to be found at this site include those of bivalves, brachiopods and belemnites. Crinoids are common enough but are usually fragmentary. Despite a low yield of the better specimens, Quantoxhead still gives plenty of opportunity for a family fossil hunt, on an impressive stretch of coast. There are always bivalves and the more common fossils to be found and the rock pools are plentiful and good to explore.

Typical ammonite found among the rocks on the foreshore at Quantoxhead.

Site summary

⚠️ Rock falls

⚠️ Slippery rocks

★★★ Accessibility

★★★★ Child suitability

★★★★ Find frequency

SSSI Site. No hammering or digging into the cliffs.

Keep away from base of cliffs.

Directions: Follow the A39 to East Quantoxhead and turn into Frog Street. Park in the car park of the Church of St. Mary, leaving a small donation. From here, walk some distance to the steps leading onto the beach.

Postcode: TA5 1EJ

CROSS HANDS QUARRY
LITTLE COMPTON, WARWICKSHIRE

Cross Hands Quarry is 5 km west of Chipping Norton. It is a great place for a family visit and is regularly visited by school parties, as the quarry allows collecting in a designated spoil heap which is regularly topped up with fresh material.

The site exposes limestone and marl belonging to the Middle Jurassic Period, encompassing the boundary between the Upper Bajocian of the Inferior Oolite and the Lower Bathonian of the overlying Great Oolite Group. The rocks are highly fossiliferous and were formerly extracted at the quarry before its closure. The quarry itself has SSSI status but this only affects the in situ bedrock and not the spoil heaps. The main quarry is not accessible to the public. Access to the spoil heaps is by prior permission.

- JURASSIC
  - BATHONIAN — GREAT OOLITE GROUP — CHIPPING NORTON LIMESTONE FORMATION
  - BAJOCIAN — CLYPEUS GRIT MEMBER

The site has some of the most extensive outcrops of the Bajocian Clypeus Grit laid down during the Middle Jurassic Period. These limestones underlie the basal units of the Hook Norton Beds. The Clypeus Grit is here seen in its most northerly exposure and the site is thus of considerable value in any palaeogeographical and palaeoecological reconstruction for this part of the Middle Jurassic.

Fossil hunting at Cross Hands Quarry requires little in the way of equipment. It is a hands and knees operation, where fossils within the spoil tips can be picked up by hand and the only tool that might be of use is a small hand trowel, to turn over the rocks.

The fossils are easy to spot and mostly consist of a good range of brachiopods (e.g. *Stiphrothyris* sp., *Acanthothyris* sp.), small echinoids (particularly *Clypeus ploti* [shown below], *Nucleolites* sp., *Holectypus* sp.), bivalves (e.g. *Pholadomya* sp., *Gresslya* sp., *Trigonia* sp.), worms, trace fossils, gastropods, plants (e.g. ginkgo), corals, bryozoans and small fragments of dinosaur bone!

The original quarry faces at Cross Hands are now protected SSSI status sites. In line with other sites, Jurassic strata were extensively quarried in Warwickshire, during the nineteenth and early to mid twentieth centuries, but much of the succession is now poorly exposed.

Site summary

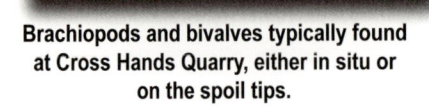

★★★★★ Accessibility

★★★★★ Child suitability

★★★★★ Find frequency

A very safe site, ideal for children and school parties.

Prior permission is required from RA Newman & Sons (Tel: 01608 674288).

Directions: From Warwick, Stratford, etc., go to Shipston on the A3400, continue to the village of Long Compton. Turn right to Great Rollright and Little Compton. Turn right onto the A44. Immediately past the Cross Hands Inn, turn right into the quarry through the first large gate. Bear right and park near the disused weighbridge office about 50 yards from the gate.

Nearest postcode: GL56 0SP (Cross Hands Inn)

Brachiopods and bivalves typically found at Cross Hands Quarry, either in situ or on the spoil tips.

Cruxicheiros newmanorum (meaning 'cross hand') is a theropod dinosaur, the first fossils of which, (pieces of scapula, hip and leg bones and vertebrae) were discovered in Cross Hand Quarry in the early 1960s. *Cruxicheiros* probably grew to around 20 feet in length.

SEDBURY CLIFF, GLOUCESTERSHIRE

SSSI Site – no hammering or digging in the cliffs

Sedbury Cliff presents much the same collecting opportunities as found at Aust, which is opposite, on the other side of the Severn. The emphasis at Sedbury, however, is on the Jurassic Blue Lias Formation in the uppermost layer of the cliff, rather than the Triassic rocks beneath, which are less frequently found. Certainly, the Liassic Beds are thicker here than at Aust, particularly at the end and middle sections of the cliff and fallen blocks are common. These are frequently full of ammonites and bivalves.

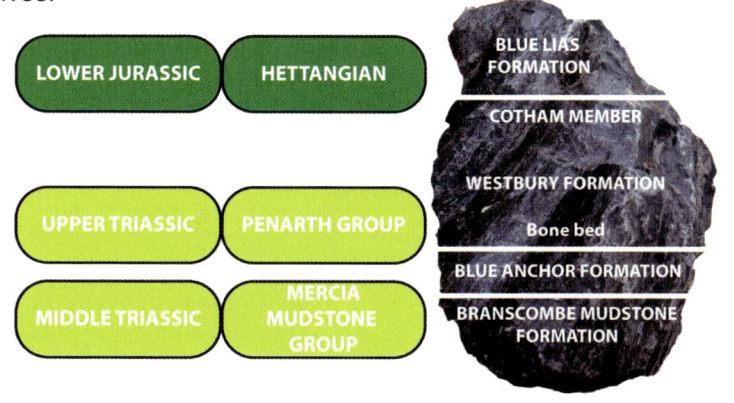

LOWER JURASSIC	HETTANGIAN
UPPER TRIASSIC	PENARTH GROUP
MIDDLE TRIASSIC	MERCIA MUDSTONE GROUP

- BLUE LIAS FORMATION
- COTHAM MEMBER
- WESTBURY FORMATION
  - Bone bed
- BLUE ANCHOR FORMATION
- BRANSCOMBE MUDSTONE FORMATION

Ichthyosaur vertebra from Sedbury, found by Damian Bartlett of DinoDeals.co.uk and used with his kind permission.

The Jurassic Lias usually falls as large slabs, with plenty of shells and poorly preserved ammonites.

Sedbury Cliff is less frequented than Aust, with overgrown vegetation in many sections but that should not deter the collector from visiting. The falls from the cliff are frequent here and can provide a good collection, especially of ammonites and bivalves.

However, it is not easy to access Sedbury Beach after bad weather. The field leading to the beach is notoriously muddy, especially just before the cliff, where the mud can be at knee height, so beware! Keep to the riverside in such conditions and wear tight fitting wellies, as the suction will leave you without footwear! Also remember that walking back to the car, with a heavy rucksack of fossils and rocks is not without difficulty in the extreme mud underfoot.

In the past, vertebrate remains were commonplace here but not so now. Ichthyosaur rib pieces and vertebrae are found but usually get quickly collected by the regular visitors and their finds invariably find their way onto eBay! Nontheless, Sedbury Cliff has much to offer by way of other finds.

Sedbury is often considered to be a 'poor relative' of Aust but the truth is that Sedbury offers more Lias material and less Triassic Bone Bed. Sedbury is never as busy either, which can only increase the possibility of finds occurring and a good day's collecting.

The foreshore at Sedbury Cliff, which can get extremely muddy.

Bivalve cluster.

Site summary

Rock falls

Slippery rocks

Deep mud

★★★ Accessibility

★★★ Child suitability

★★★★ Find frequency

SSSI site. No hammering or digging in the cliffs.

Beware of very muddy conditions leading to the beach from adjoining fields.

Directions: From Chepstow, take the road to Sedbury. Park at the lay-by outside Three Salmons House at NP16 7HG. Walk back towards the footpath marked to Offas Dyke, then the footpath marked to Sedbury Cliffs. Walk is about 1 mile to get to the site.

Postcode: NP16 7HG

FOLKESTONE, KENT

SSSI Site – no hammering or digging into the cliffs

This highly fossiliferous site, especially on the foreshore near Copt Point, is composed of rocks from the Lower Cretaceous; Gault Clay and Lower Greensand. The rocks are of the Middle Albian stage, at Folkestone being approximately 106 million-years-old. At the base of the Gault Clay cliffs is the Lower Greensand Formation. Boulders from this are strewn over the foreshore, making it hard walking in some sections.

The dark bluish-grey Gault Clay and mudstones have slipped over the cliff section of Greensand but its rich yield of fossils can be found in situ and washed out of the clay by the sea, to be deposited between the boulders. The Lower Greensand is from the Early Albian stage and is not as fossiliferous. It is the Gault Clay for which Folkestone is famous, especially for its iridescent ammonite shells.

CRETACEOUS

LATE ALBIAN — UPPER GREENSAND

MIDDLE ALBIAN — GAULT CLAY

EARLY ALBIAN — LOWER GREENSAND

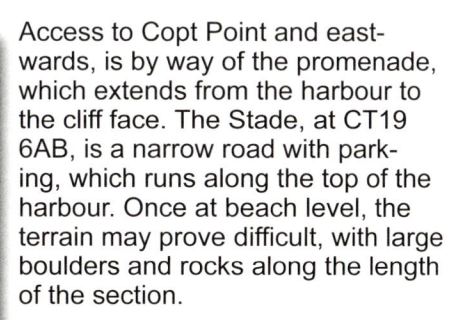

Access to Copt Point and eastwards, is by way of the promenade, which extends from the harbour to the cliff face. The Stade, at CT19 6AB, is a narrow road with parking, which runs along the top of the harbour. Once at beach level, the terrain may prove difficult, with large boulders and rocks along the length of the section.

The Lower Greensand contains mostly thick-shelled molluscs but the Gault Clay is the source of a number of fossil types, including fish, turtles, crabs, lobsters, gastropods, bivalves, shark teeth, belemnites and ammonites. The sheer variety and frequency of finds makes Folkestone a highly productive site.

Fossils can be found in situ in the base of the cliff, among the boulders and rocks strewn along the foreshore and during scouring conditions, within the Gault Clay itself, exposed on the foreshore.

The site has SSSI status, so digging into the cliff or exposures on the foreshore is strictly prohibited. However, collection of many fossils is easily possible through collection by hand, where the specimens protrude from the clay after erosion by the sea.

Specimen collection boxes or a suitable container with tissue is advised for collection purposes. It is quite probable that the collector will obtain a high number of finds from this locality, given the right conditions. Collection after a storm or during winter months is always more productive.

The cliffs and foreshore at Copt Point, Folkestone - a mixture of Gault Clay with Upper Greensand boulders. Fossils shown include *Euhoplites* ammonite, *Birostrina sulcata* and an ichthyosaur tooth.

The staggering number of types and the beauty of the ammonites found at Copt Point, Folkestone, are undoubtedly the reason for the popularity of this location for the collector. Top row (left to right): *Euhoplites alphalautus, Euhoplites trapezoidalis, Anahoplites planus, Anahoplites aff. splendens*. Bottom row (left to right): *Dimorphoplites parkinsoni, Hoplites dentatus, Euhoplites inornatus, Mortoniceras inflatum*.

A rare turtle skull, found by UKAFH member Phil Hadland.

Highly iridescent ammonites are a common find.

An assortment of shark teeth from the Gault Clay.

Site summary

⚠️ Rock falls

⚠️ Slippery rocks

★★ Accessibility

★★ Child suitability

★★★★ Find frequency

SSSI Site. No hammering or digging into the cliffs or bedrock.

Crumbling clay cliffs and rocky foreshore make this a site for older children only.
Access to foreshore is difficult.

Suggested equipment: Steel point, specimen collection box, wrapping material (cotton wool, tissue).

Directions: Park on The Stade, walk along the promenade eastwards and down onto the foreshore near Copt Point.

Postcode: CT19 6AB

DANES DYKE, EAST RIDING OF YORKSHIRE

SSSI Site – no hammering or digging into the cliffs

The Chalk at Dane's Dyke is famous for its abundant fossils, especially echinoids and sponges. The Late Cretaceous rocks are of Campanian age (83–71 mya) and the South Landing Member, the Danes Dyke Member and the Sewerby Member of the Flamborough Chalk Formation are all present. The older Sewerby Member is prominent in the western end of the cliff, whilst the older South Landing Member crops out at the eastern end.

The Chalk at this location is soft, so extraction and preparation of any fossils collected is easy. Other fossils to be found here include bivalves, belemnites and crinoids.

FLAMBOROUGH CHALK FORMATION

CRETACEOUS **CAMPANIAN**

SOUTH LANDING MEMBER

DANES DYKE MEMBER

SEWERBY MEMBER

The vast numbers of boulders and fallen rocks at Danes Dyke present a great opportunity to find fossils within. It is just a matter of splitting the rock, using a bolster chisel and geological hammer. The location is well known for the Chalk sponges, which can be large and unusual. The better sponges are found using the method described and are not weathered or eroded. Searching among the rock debris will also yield echinoids and crinoids, especially the plates.

The Chalk cliffs are free of flint, with noticeable bands of marl present. Fossils are found in the cliff but it is an SSSI site so you must confine any fossil collecting to the beach. Avoid searching at the base of the crumbling cliffs and besides, more productive results will be obtained from foreshore collecting in the fallen material. The regular supply of fresh material from falls ensures that there is plenty to find.

The rocky Chalk foreshore at Danes Dyke is highly fossiliferous and new cliff falls bring regular fresh material to the beach.

Site summary

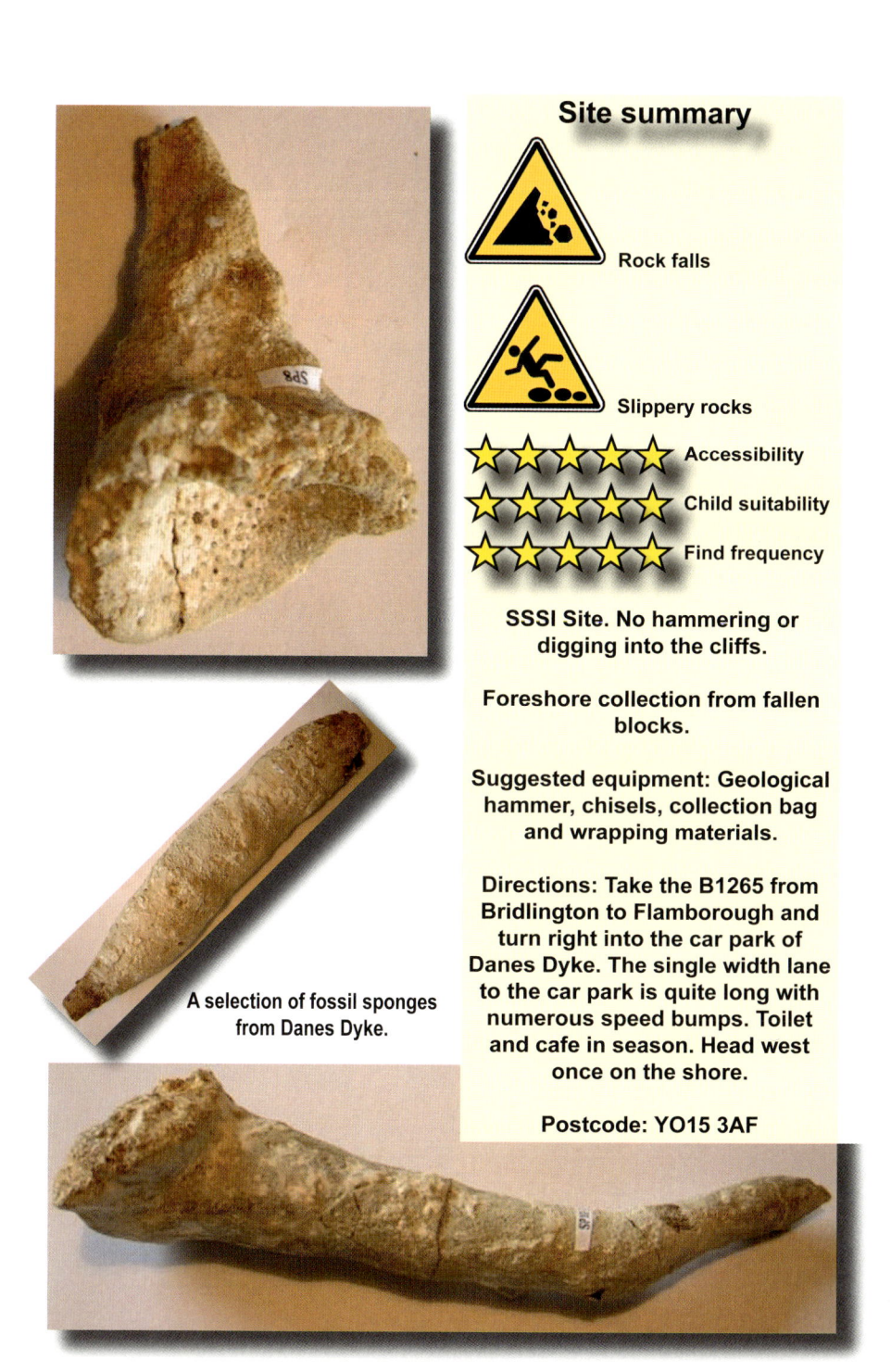

⚠️ Rock falls

⚠️ Slippery rocks

★★★★★ Accessibility

★★★★★ Child suitability

★★★★★ Find frequency

SSSI Site. No hammering or digging into the cliffs.

Foreshore collection from fallen blocks.

Suggested equipment: Geological hammer, chisels, collection bag and wrapping materials.

Directions: Take the B1265 from Bridlington to Flamborough and turn right into the car park of Danes Dyke. The single width lane to the car park is quite long with numerous speed bumps. Toilet and cafe in season. Head west once on the shore.

Postcode: YO15 3AF

A selection of fossil sponges from Danes Dyke.

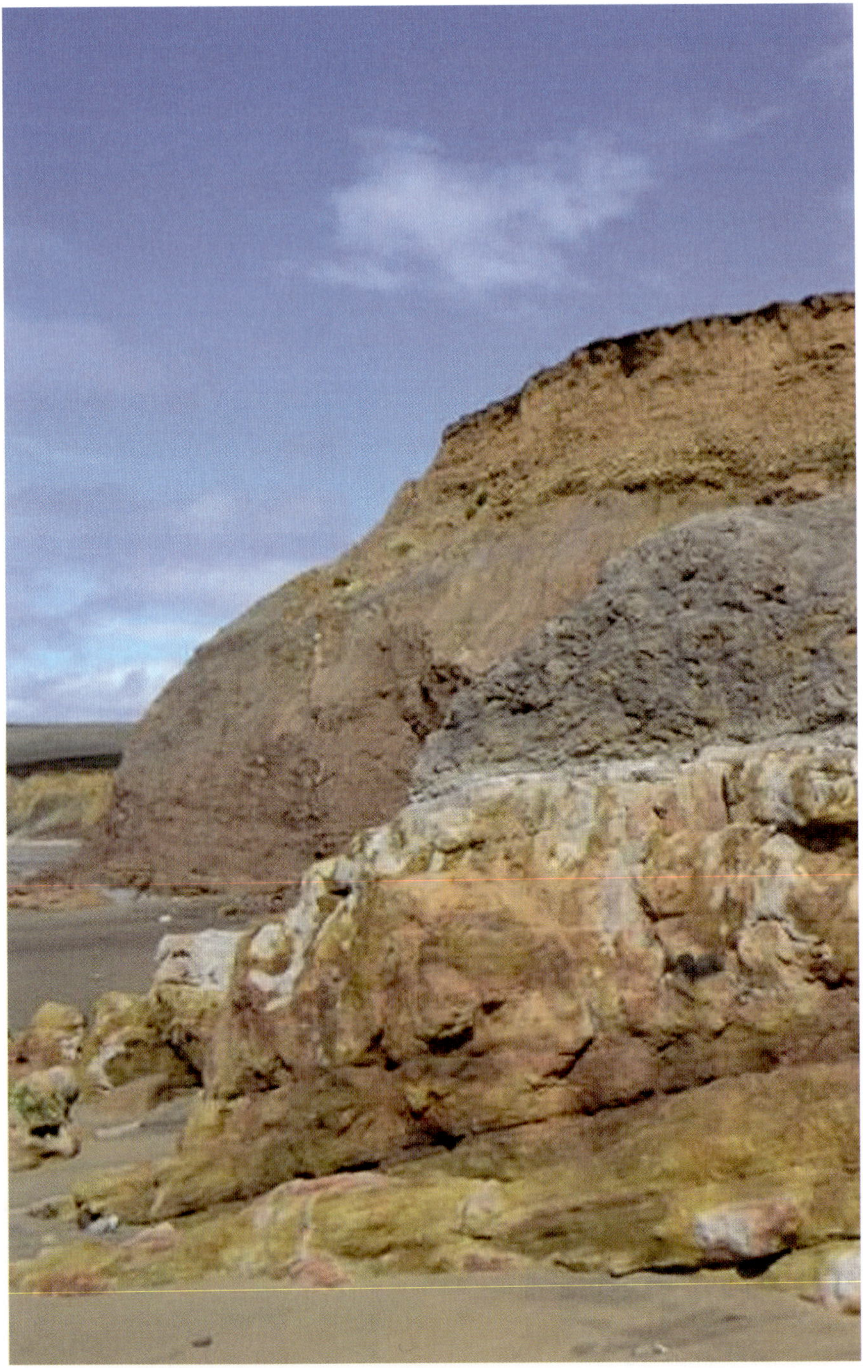

COMPTON BAY & BROOK BAY, ISLE OF WIGHT

SSSI Site – no hammering or digging into the cliffs or bedrock

Compton Bay and nearby Brook Bay, on the southwest coast of the Isle of Wight, are places of rapid coastal erosion. The rocks are all Cretaceous at this location and, in the main, belong to the Lower Cretaceous Wealden Group of the Barremian stage (129.4–125 mya). The Chalk is equally well exposed and sandwiched between are exposures of the Lower Greensand, Gault and Upper Greensand.

This section is well known for the remains and footprints of dinosaurs, for which the Isle of Wight is famous. Both commonly occur on the foreshore. In line with regulations on the island, the collection of dinosaur footcasts is strictly forbidden.

CRETACEOUS   BARREMIAN   WEALDEN   WESSEX FORMATION

Iguanodontid footprint at Compton Bay.

Brook Bay.

Access to the beach at Compton Bay is easy, with a car park situated at Compton Chine.

Hanover Point is a short distance southeast of Compton Bay and is probably the most well known of the dinosaur fossil localities, with both bones and footprints present. The stretch of beach from Compton Chine to Brook Chine is easily accessed from the car parks located at either end. However, the site is protected by the National Trust, preventing the removal of the large, dinosaur footprints.

Some are still embedded in the source exposure, a white splay crevasse sandstone, which extends from just west of Hanover Point to about 100 metres west of Brook Chine. They are well exposed for 15 metres at the western end. There is also a section of dinosaur trackway in a red clay bed, 150 metres out from the cliff at Hanover Point, heading in a southeasterly direction, but these are only accessible at a low tide.

The entire section of coast along Compton Bay and through Brook Bay provides a worthwhile day, with opportunities to find pieces of dinosaur bone (mostly rolled) and possibly teeth, in addition to seeing the large number of dinosaur footprints scattered along the coast.

Brook Bay is the site of the oldest Cretaceous rocks and fossils exposed on the Isle of Wight. The purple, blue and pink sediments of the cliffs and foreshore were deposited on an ancient river floodplain around 126 million years ago.

Dinosaurs like the giant *Iguanodon*, armoured *Polacanthus foxii* (above) and the rare meat-eaters walked this ancient landscape and left behind evidence as foot casts and tracks.

Rolled dinosaur bone from Compton Bay.

Vertebra of a dinosaur.

Large footcast, possibly *Mantellisaurus*.

Site summary

⚠️ Rock falls

⚠️ Slippery rocks

★★★★★ Accessibility

★★★★★ Child suitability

★★★★ Find frequency

SSSI Site. No hammering or digging into the cliffs or bedrock. Dinosaur footprints must not be removed.

Suggested equipment: Lump hammer or geological hammer, chisels, specimen collecting bag, small hand pick.

Directions: You can park at the Compton Beach car park, with easy access down to the shore from the cliff top car park. Parking is also available at Brook Chine car park, at the southern end of the bay.

Postcodes: Brook Chine National Trust car park, Military Road, Newport PO30 4HA, Compton Beach car park PO30 4HB

YAVERLAND, ISLE OF WIGHT

SSSI Site – no hammering or digging into the cliffs

Yaverland is situated on the southeast side of the Isle of Wight, between Bembridge and Shanklin. This Cretaceous site exposes the Wealden beds of the Wessex Formation and Vectis Formation, with later exposures of Upper Greensand, Gault and Chalk.

In the main, it is a highly fossiliferous site, famed for the bones and teeth of dinosaurs, crocodiles, turtles and fish, as well as plant fossils. In reality, Yaverland can sometimes be disappointing if the scouring tidal conditions are not favourable and the site can certainly be over-collected. Nonetheless, most people will find something at this location and the foreshore will provide a good sandy beach, ideal for younger children and families.

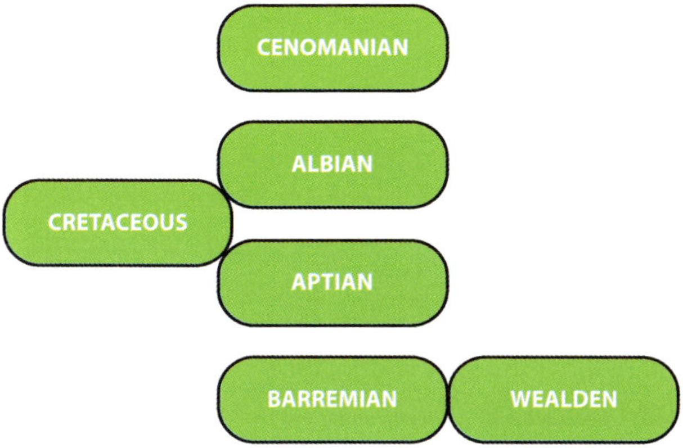
CENOMANIAN
ALBIAN
CRETACEOUS
APTIAN
BARREMIAN
WEALDEN

CHALK
UPPER GREENSAND
GAULT
CARSTONE
SANDROCK
FERRUGINOUS SANDS
ATHERFIELD CLAY
VECTIS FORMATION
WESSEX Fm.

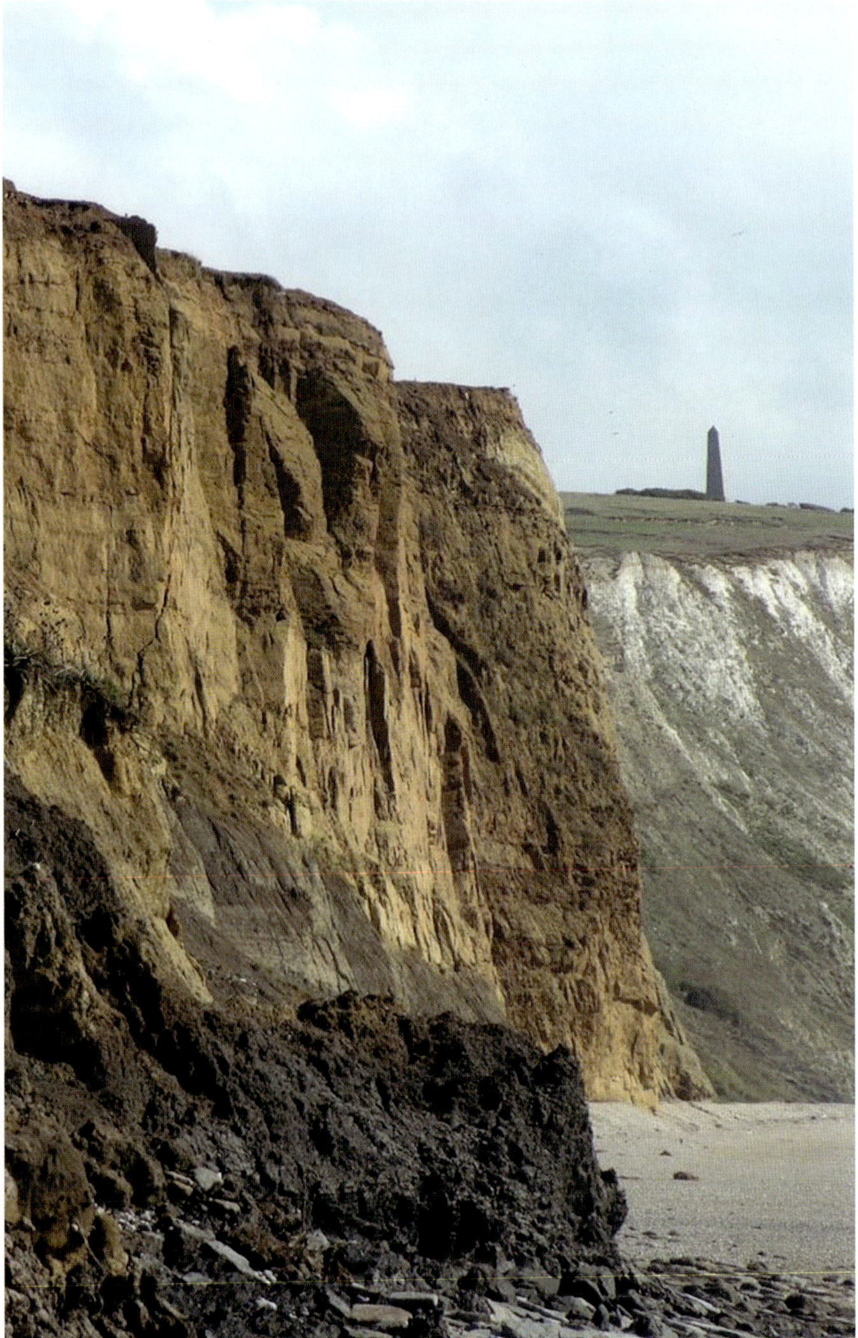

At Yaverland, the cliff starts with Barremian-aged rocks (127–121 mya) of the Wessex Formation within the Wealden Group. Wealden exposures continue from this into the Vectis Formation. Here it is possible to find mudstone blocks full of clams. Fish scales and shark spines can also be found in these layers.

Following on from this, the Atherfield Clay Formation of Aptian age (121– 111 mya) can be seen. This contains a hard limestone layer, which can be found in blocks exposed on the beach.

At the far end of the bay, the Ferruginous Sands can be seen in the distance marked by the orangey red coloured cliff, followed by Carstone, Gault and then the Upper Greensand. Finally, the headland is marked by Chalk.

Fossil collecting at Yaverland is very carefully monitored and collectors are not allowed to remove specimens from the beds or foreshore which are larger than 1 foot (30 cm) across. Larger, important finds should be reported to the appropriate authority.

UKAFH field event in the Wealden rocks at Yaverland in 2015.

The Ferruginous Sands in the distinctive orange-red cliffs at Yaverland.

Atherfield Clay slumped onto the beach at Yaverland.

Vertebra of *Polacanthus foxii*, an armoured dinosaur, found by Craig Chapman at Yaverland.

Rolled dinosaur bone.

Site summary

⚠️ Rock falls

★★★★★ Accessibility

★★★★★ Child suitability

★★★ Find frequency

SSSI Site. No hammering or digging into the cliffs.

Collection of fossils from loose material on the foreshore and beach. Stay clear of cliff base. Removal of material larger than 30 cm is not allowed. Car parking and toilets close by.

Suggested equipment: Lump hammer or geological hammer, chisels, specimen collecting bag, small hand pick.

Directions: From Sandown, follow the B3395 east to Yaverland. There is easy access to the beach, with a large car park near the shore.

Postcode: PO36 8QB

EASTBOURNE, EAST SUSSEX

SSSI Site – no hammering or digging into the cliffs

The Upper Cretaceous Chalk at Eastbourne is highly fossiliferous and considered to be one of the best Chalk sites in the UK. The erosional rate is fast, so fresh material is constantly hitting the beach from the cliffs above, with good yields of fossils as a result.

Eastbourne and the surrounding South Downs and Chalk cliffs reveal a marine palaeoenvironment, which at Eastbourne dates to 86 Ma. Fossils occur commonly throughout the highly fossiliferous chalk, including ammonites, echinoids, bivalves, bryozoans, sponges and corals.

- CRETACEOUS
  - WHITE CHALK SUBGROUP
    - LEWES NODULAR CHALK FORMATION
    - NEW PIT CHALK Fm.
    - HOLYWELL NODULAR CHALK FORMATION
  - GREY CHALK SUBGROUP
    - ZIG ZAG CHALK FORMATION
    - WEST MELBURY MARLY CHALK Fm.

UKAFH fossil hunt at Eastbourne, March 2016.

Eastbourne foreshore looking east.

Eastbourne beach and cliffs towards Cow Gap.

Fossils can be found in and amongst the large chalk blocks and boulders on the foreshore and towards the cliff base. Please do not hammer into the cliff face. In any case, working directly under the cliffs is dangerous, as they can, and do, collapse. Using a wide chisel and hammer, carefully cut around the fossil. Splitting the blocks can also reveal fossils. Work westwards along the beach from beyond the groynes to Cow Gap and towards Beachy Head.

At Eastbourne, most fossils are found in the section high in the cliffs, so looking among the fallen rocks and foreshore is the best method of collecting fossils.

Chalk fossils need to be desalinated and washing over for several days in clean water and then treating with an application of a coat or two of PVA, diluted 1:3 with water will stabilise and preserve them. Chalk fossils will disintegrate in time if left untreated. An easy way to desalinate smaller fossils, like shells and echinoids, is to place them in an old stocking or tights and drop the 'bag' into the toilet cistern. This will ensure a regular change of fresh water after each flush. After a week, the fossils can be naturally left to dry and then treated as suggested. Larger fossils, such as ammonites, will need to be desalinated in a container of water, changed regularly.

Cenoceras, a nautilus found west of Holywell by UKAFH member Chris Strutt.

A selection of fossils found at Eastbourne in 2014 and 2016, by members of UKAFH, on an organised field trip. Left is a *Spondylus spinosus* bivalve, right is a sponge.

Mantelliceras ammonite found by Chris Strutt.

Mantelliceras found by Chris Strutt.

Micraster echinoids.

Acanthoceras rhotomagense ammonite found and prepared by Chris Strutt.

The section between Holywell to Cow Gap and onwards to Beachy Head will require two visits. It is a long stretch of coast and for first time visitors or families, the stretch of low-gradient cliff immediately southwest of the steps at Cow Gap provides a relatively safe and productive location to search for fossils.

The chalk here belongs to the West Melbury Marly Chalk Formation (of the former Lower Chalk) and is rich in ammonites and bivalves in particular. Further west, the Zig Zag Chalk Formation crops out and eventually the Holywell Nodular Chalk Formation and New Pit Chalk Formation (of the previously named Middle Chalk) come down to beach level.

At Eastbourne, it is a firm white chalk, with nodules and flint seams. Fossils may be abundant, especially brachiopods *Terebratulina* and *Gibbithyris* and bivalves.

The Chalk here is distinguished by its light grey colour and may contain ammonites, such as *Mantelliceras*. Echinoids *Sternotaxis*, *Micraster* and *Echinocorys* are common and can often be found on the beach, either as specimens that have washed out of their matrix or as flint casts. The crinoid *Marsupites* and the small sponge *Porosphaera* are also found, along with various corals.

Site summary

Rock falls

Slippery rocks

★★★★ Accessibility

★★★ Child suitability

★★★★★ Find frequency

SSSI Site. No hammering or digging into the cliffs or bedrock.

Rocky foreshore can be slippery.

Suggested equipment: Geological hammer, chisels, goggles, wrapping material.

Directions: Follow the B2103 out of Eastbourne heading west along the sea front road, which is King Edward's Parade. This section of road (heading towards Beachy Head) eventually becomes Dukes Drive. This is the Hollywell area of the town. There is free parking along this entire stretch of road or opposite in Chesterfield Road. The free parking on the sea front section is beside a grassed area, overlooking the beach from above. Once parked make your way down the various steps and pathways towards the beach area and cliffs.

Postcode: BN20 7NU

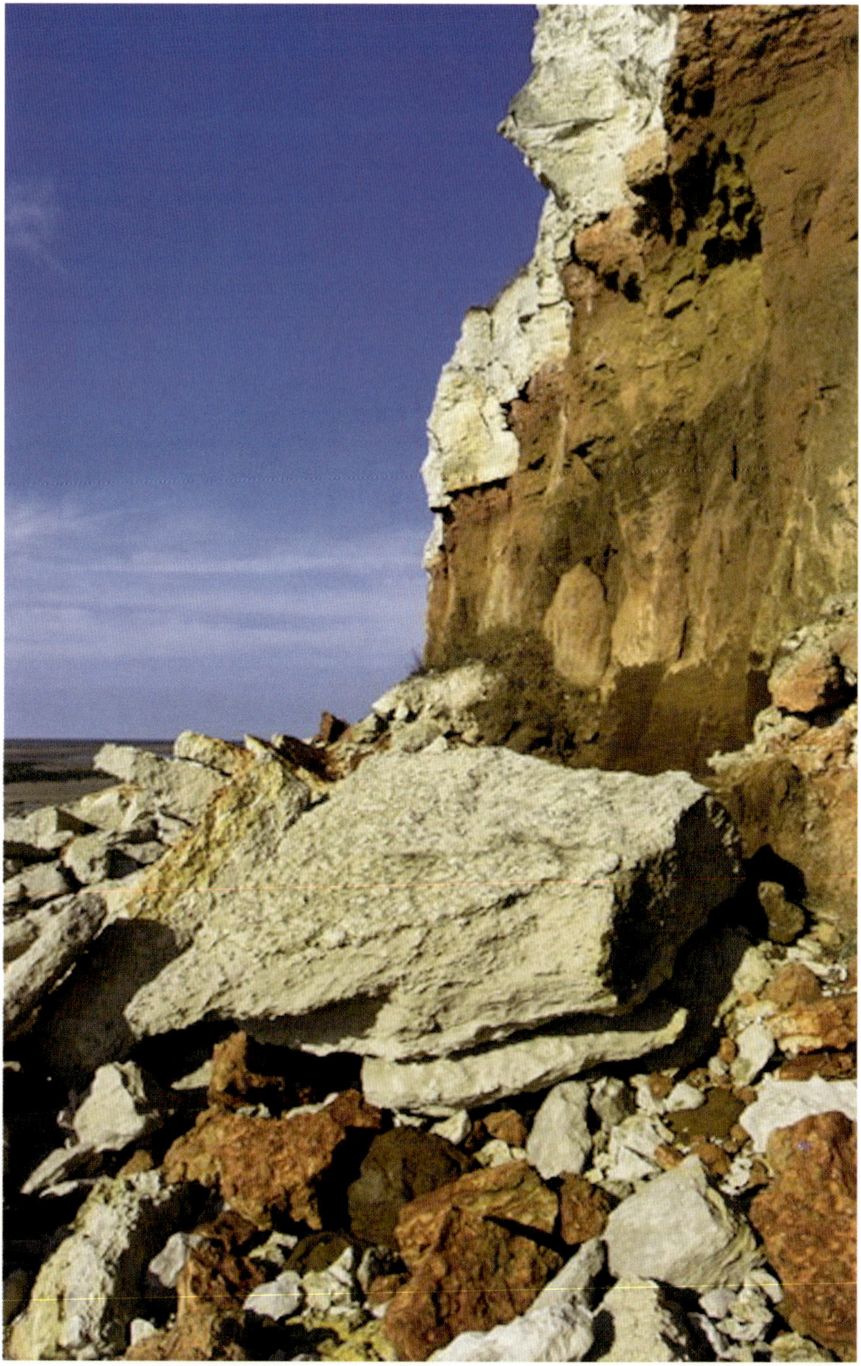

HUNSTANTON, NORFOLK

SSSI Site – no hammering or digging into the cliffs

The impressive colour contrasting cliffs of orange, red and white sedimentary rocks at Hunstanton, on the northwest coast of Norfolk, reflect changing depositional conditions towards the end of the Early Cretaceous and the onset of the Late Cretaceous.

The rocks are of marine origin and date from the Albian (108 Ma) to Cenomanian (99 Ma). The orange Carstone Formation at the base is overlaid by the red Hunstanton Formation (also known as Red Chalk) and finally the white Ferriby Chalk, which extends to the top of the cliff.

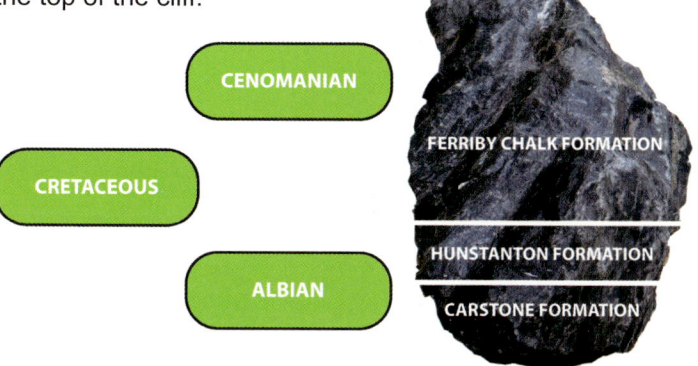

CENOMANIAN

CRETACEOUS

ALBIAN

FERRIBY CHALK FORMATION

HUNSTANTON FORMATION

CARSTONE FORMATION

The basal Carstone Formation is a sandstone, composed of coarse sand particles, with rolled pebbles and fossils present. The Hunstanton Formation is, by comparison, a one metre-thick bed of red coloured limestone, in which brachiopods, belemnites, corals and echinoids occur. The Ferriby Chalk Formation is high above the cliff face and fossil collection from it is reliant on the frequent cliff falls and therefore collection at beach level. The use of a geological hammer and chisels will be required to split the Chalk rocks and to extract specimens revealed in the beach debris.

The cliffs and foreshore at Hunstanton.

Using a wide chisel and hammer, carefully cut around the fossil. Finer preparation can be dealt with back at home, using methods described in an earlier chapter.

Fossil collection at Hunstanton can be very productive and fossils that might be found here include ammonites, belemnites, echinoids, brachiopods, bivalves, sponges, worm tubes, corals and crustacean burrows.

Photo ©Richard Humphrey and licensed for reuse under Creative Commons License.

Rock falls are a common occurrence, so finding fresh material on the beach for searching through will present no problem. Hunstanton is a good place for a family fossil hunt, with a high frequency of finds and an interesting coastline.

The constant erosion of the cliffs at Hunstanton causes substantial falls, which enable collection from fallen blocks on the beach. Photo ©Peter Pearson and licensed for reuse under Creative Commons License.

Echinoid fossil from the Carstone Formation.

Sponge protruding from a foreshore boulder.

Belemnite in the Carstone Formation.

Site summary

⚠️ Rock falls

⚠️ Slippery rocks

★★★★ Accessibility

★★★★★ Child suitability

★★★★★ Find frequency

SSSI Site. No hammering or digging into the cliffs or bedrock.

Fossil collection from fallen blocks on the beach.

Suggested equipment: Geological hammer, chisels, goggles, specimen collection bags, wrapping materials.

Directions: Access to the beach is made at St. Edmond's Point a short distance north of the ruins of St. Edmond's Chapel. Walk north along the cliff top footpath to the beach access point.

Postcode: PE36 6EL

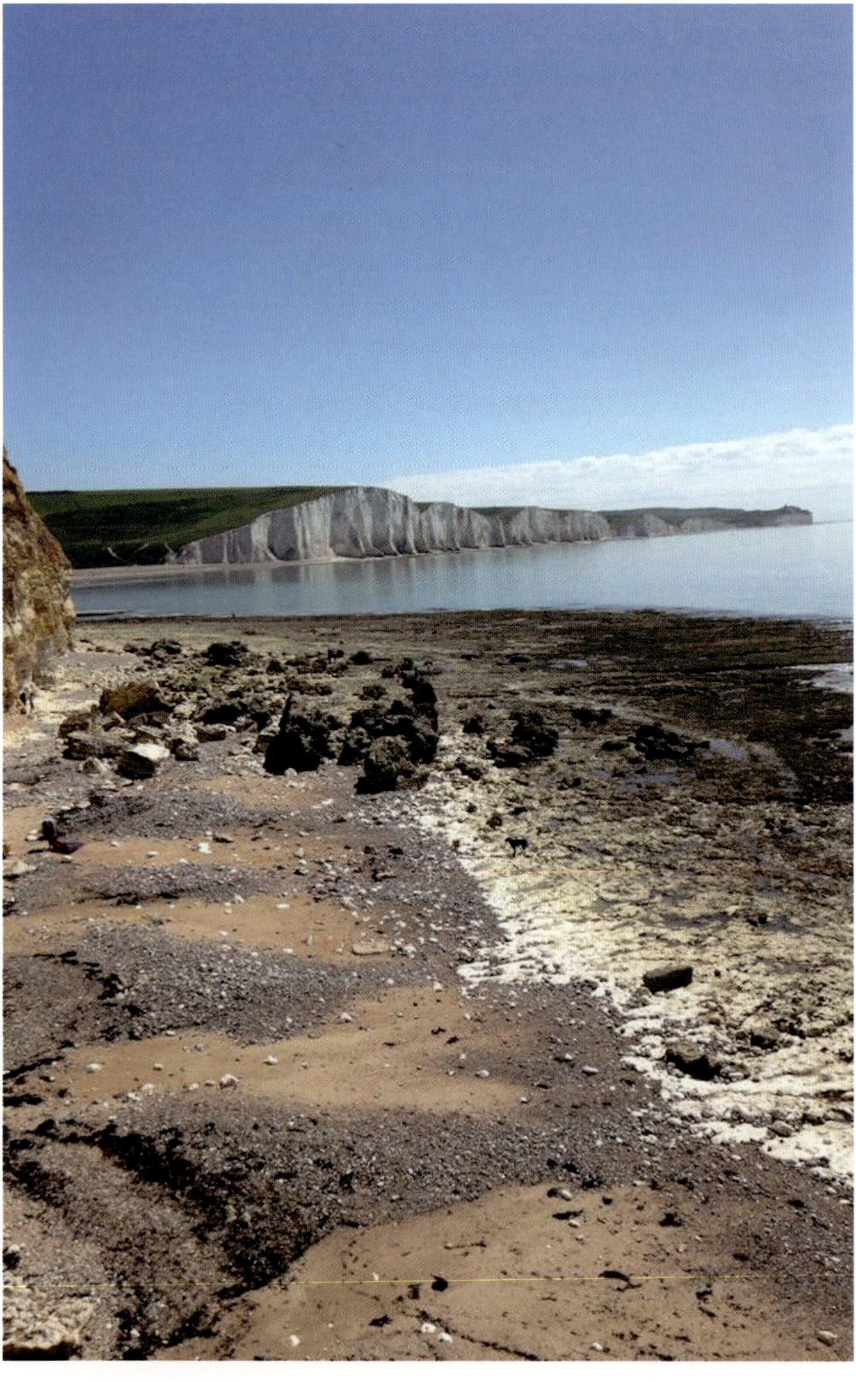

SEAFORD, EAST SUSSEX

SSSI Site – no hammering or digging into the cliffs or bedrock

The late Cretaceous Chalk cliffs and foreshore at Seaford Head provide a good site for the Chalk fossil enthusiast, with plenty of echinoids, sponges, bivalves, and other benthic fauna that inhabited the seafloor during the Coniacian stage (approximately 89–86 mya).

The earliest chalk at Seaford Head belongs to the Lewes Nodular Chalk Formation and dates from 89 Ma. The formation is well exposed on the foreshore and in the lower half of the cliff. Above this, the Seaford Chalk Formation appears higher in the cliff.

CRETACEOUS   CONIACIAN   WHITE CHALK SUBGROUP

SEAFORD CHALK FORMATION

LEWES NODULAR CHALK FORMATION

Hope Gap Sheet Flint

From the steps at Hope Gap towards Seaford, an intermittent ledge protrudes from the cliff base. The upper surface indicates the top of the Hope Gap Hardground; a conspicuous layer composed of iron-stained nodular white chalk, interspersed with soft, grey chalk.

Above the Hope Gap Hardground are the Beeding Hardgrounds, also belonging to the Lewes Nodular Chalk Formation, containing an abundance of in situ shattered flints.

Fossils can be found in any direction once on the beach at Hope Gap, although the better exposures are located towards Seaford. The foreshore is certainly the best and most productive area and is safely away from the cliff face.

Chalk boulders, rocks and flint nodules from the cliffs are strewn across the foreshore and provide a constant supply of fossils. The cliffs and platform are designated SSSI, so any fossil collecting efforts must be directed at loose boulders and rocks on the foreshore.

A common find, a *Micraster* echinoid in flint.

A regular *Temnocidaris* echinoid exposed in situ on the surface of the chalk. Image above and below used with permission of Discovering Fossils.

A complete *Spondylus* bivalve visible on both sides of a flint nodule.

Site summary

Rock falls

Slippery rocks

★★★★☆ Accessibility

★★★★☆ Child suitability

★★★★★ Find frequency

SSSI Site. No hammering or digging in the cliffs or bedrock. Rocky foreshore can be very slippery. Tidal location.

Suggested equipment: Geological hammer, chisels, goggles, wrapping material.

Directions: Immediately southeast of Seaford town is the Seaford Head Nature Reserve. It is not easy to find but drive to Chyngton Way and the road opposite Chyngton Lane is the road to the South Hill Barn car park. Access to cliffs and foreshore is through the nature reserve, at the top of which free parking is available year round. From the car park the path splits in three directions, the middle, unsurfaced route leads to a cattle grid, at which point a second path on the right leads directly to Hope Gap. At Hope Gap three flights of concrete steps extend to the beach.

Nearest postcode: BN25 4JQ

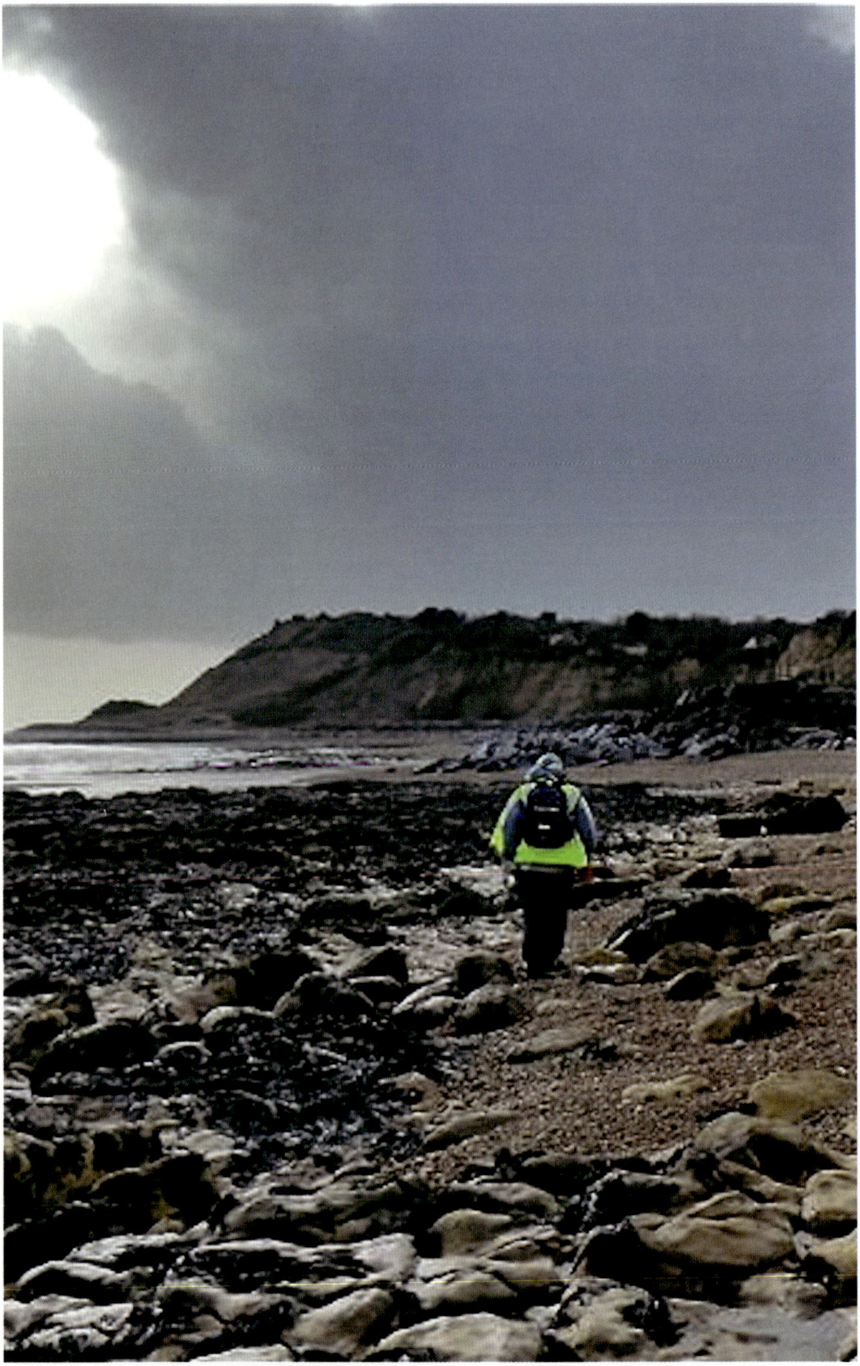

FAIRLIGHT, EAST SUSSEX

SSSI Site – no hammering or digging into the cliffs

The section of coast between Pett Level and Hastings comprises layers of clay and sandstone from the early Cretaceous. The cliffs here are unstable, yet provide the collector with much fallen material, with fossils from a freshwater lake or lagoon and foreshore that once dominated the landscape. Sands and silt were transported by rivers and streams, to form the lake bed. It is possible to find dinosaur remains here, along with footprint casts and fossil plants (especially the horsetail *Equisetites*), freshwater bivalves and fish.

The cliffs are of Berriasian (140 Ma) and Valanginian (139 Ma) age, consisting of a sandstone, the Ashdown Formation at their base, followed by Wadhurst Clay, which comprises a thin shale layer with ironstone. Above lies the Cliff End Sandstone, capped by the Wadhurst Clay Formation.

LOWER CRETACEOUS

VALANGINIAN

BERRIASIAN

WADHURST CLAY

CLIFF END SANDSTONE

WADHURST CLAY (shales+ironstone)

ASHDOWN FORMATION (Sandstone)

Access to the beach is made east of Fairlight at Pett Level, where an access road to the beach runs alongside The Smuggler pub at TN35 4EH. Park along the main road opposite the pub.

Again, we have to emphasise the instability of the cliffs here and for collectors to stay clear of the cliff base at all times. Collecting is best along the rocky foreshore, where fallen blocks should be searched, although fossils are not always evident. The Cliff End Bone Bed is evident at the start of the cliffs and rocks from this are worth splitting for bone and teeth. Everyone likes the idea of finding a dinosaur bone but theses are relatively scarce.

However, searching amongst the boulders can reveal bones and the remains of turtle or crocodile, fish teeth and scales. Bones are invariably dark brown in colour. The commonest fossil is of the bivalve *Neomiodon*, which can be seen in many rocks and boulders. Plant fossils are common also. Lignite, often referred to as brown coal, is a soft brown combustible sedimentary rock formed from naturally compressed peat and is present in some rock layers.

Dinosaur footprints are a regular occurrence and these include those made by iguanodonts, ankylosaurs and theropods.

The foreshore at Fairlight Cove looking west.

Bone bed with multiple fish, hybodont and pterosaur remains.

Rolled dinosaur bone from Fairlight.

Theropod footprint on the beach.

A crocodile block from the Bone Bed, containing some large bones and crocodile scutes.

Dinosaur footprint, possibly that of an *Iguanodon*.

Site summary

⚠️ Rock falls

⚠️ Slippery rocks

★★★ Accessibility

★ Child suitability

★★ Find frequency

SSSI Site. No hammering or digging into the cliffs or bedrock.

Tidal location. Foreshore collection. Rocky foreshore can be slippery. Stay clear of unstable cliff.

Suggested equipment: Geological hammer, chisels, goggles, wrapping material.

Directions: Take the coastal road through Fairlight and Fairlight Cove to Pett Level. Park near The Smugglers pub, Pett Level Road and walk to the beach.

Postcode: TN35 4EH

Fairlight Cove is a classic collecting area and is probably the best area for future finds of Lower Cretaceous reptiles outside the Isle of Wight. Important finds of dinosaur bones should be reported to the Natural History Museum, London.

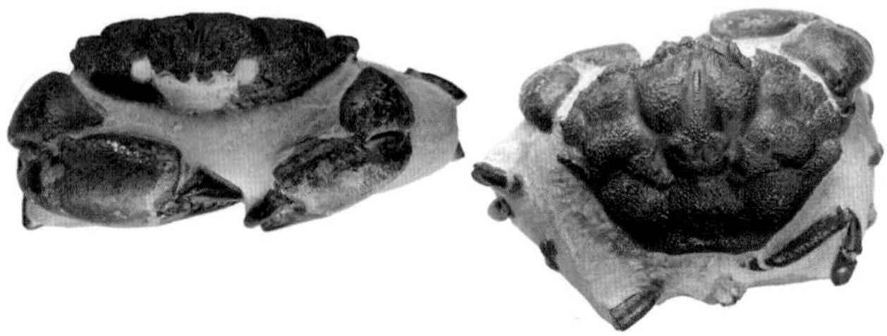

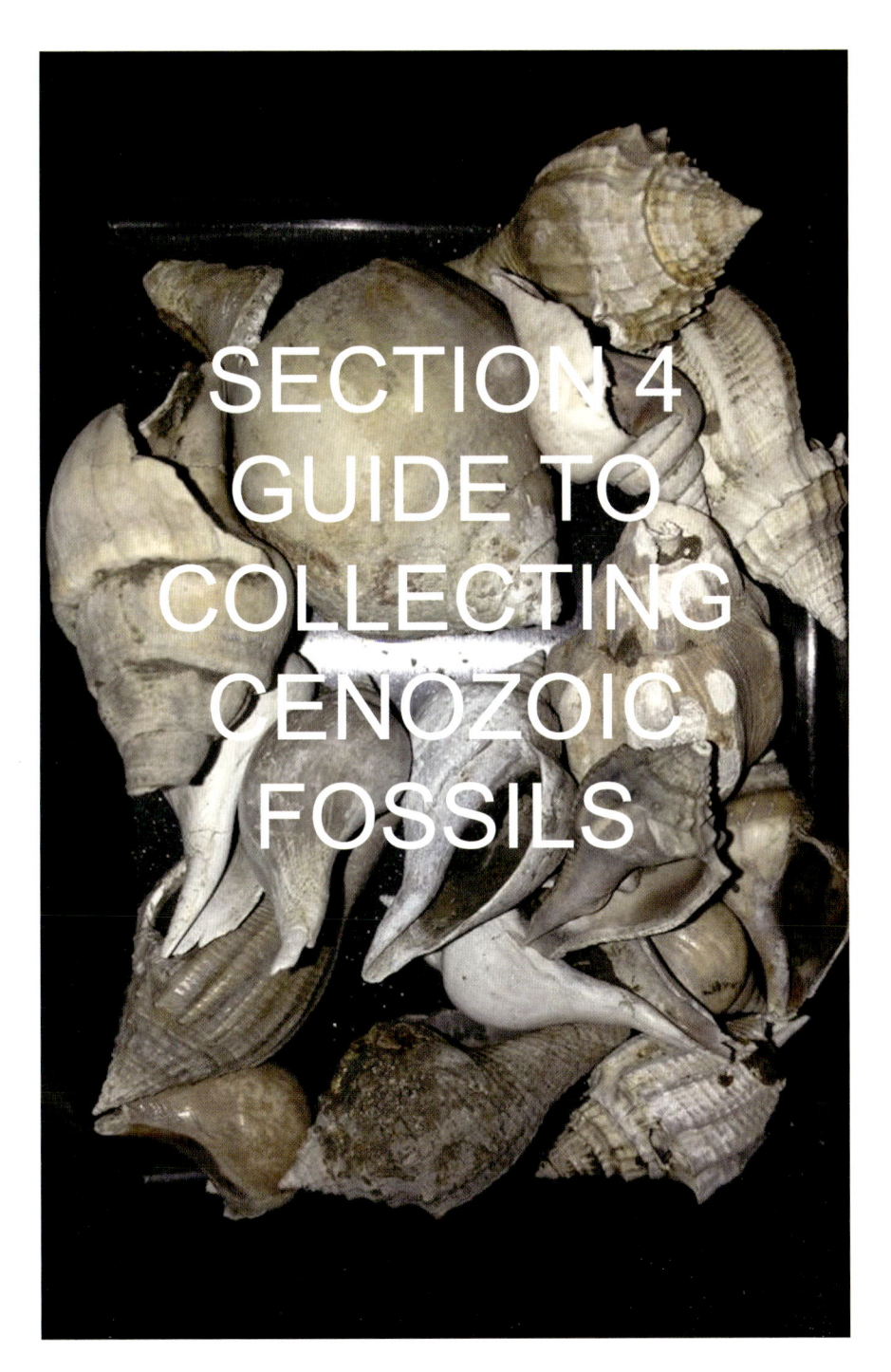

SECTION 4
GUIDE TO COLLECTING CENOZOIC FOSSILS

CENOZOIC FOSSILS

The Cenozoic era began about 65 million years ago and still continues. It is divided into three periods: the Palaeogene, Neogene and Quaternary.

The first period, the Paleogene, features three epochs: the Paleocene, the Eocene and the Oligocene. The Neogene period features two epochs: the Miocene and the Pliocene. The third period, the Quaternary features the Pleistocene and the Holocene epochs. In addition, the Anthropocene has been proposed (though not yet formally accepted) as a new epoch dating from when human activities started to have a significant global impact on Earth's geology and ecosystems.

The range and variety of life was enormous. Known as the Age of Mammals, the Cenozoic era was dominated by relatively small fauna. Small mammals, birds, reptiles and amphibians flourished and diversified, especially after the demise of the giant reptiles.

In Great Britain, the Cenozoic rocks are almost entirely confined to southeastern England, their outcrops being mainly situated in the London and the Hampshire Basins.

As with the previous sections in this book, each locality has a detailed description of the site, its geology and what you can expect to find there. Please ensure that you have checked for necessary permissions before your visit, and where applicable, check the tide times carefully.

CENOZOIC FOSSIL SITES

The following Cenozoic sites are described in this section, with the following colour code applied, as shown in the geological time chart below.

Paleocene
　Beltinge, Herne Bay, Kent
　Abbey Wood, London SE2

Eocene
　Burnham-on-Crouch, Essex
　Barton-on-Sea, Hampshire
　Minster, Isle of Sheppey, Kent
　Warden Point, Isle of Sheppey, Kent
　Maylandsea, Essex
　Bracklesham Bay, West Sussex
　Walton-on-the-Naze, Essex

Oligocene
　Thorness Bay, Isle of Wight

Pleistocene
　Easton Bavents, Suffolk
　West & East Runton, Norfolk
　Ramsholt, Suffolk
　Mappleton, East Riding of Yorkshire

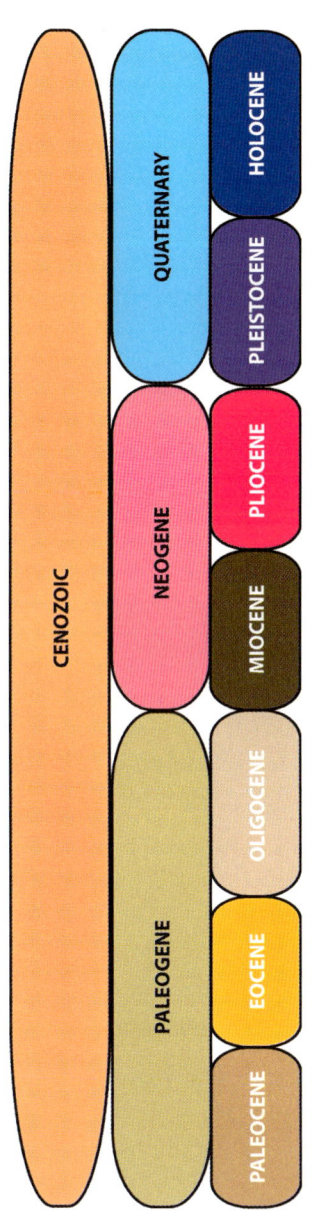

Bishopstone Glen, at Beltinge, Kent. The beach here is famed for the profusion of shark teeth to be found, which include: top row (left to right): *Notidanodon* sp., *Otodus obliquus*; middle row: *Stratiolamis macrota*; bottom row (left to right): *Palaehypotodus rutoti*, *Carchartas hopei* (formerly *Odantaspis hopei*).

BELTINGE, NEAR HERNE BAY, KENT

SSSI Site – no hammering or digging into the cliffs or bedrock

Beltinge is situated to the east of Herne Bay, on the Kent coast. The stretch from Herne Bay to Reculver provides an opportunity to collect fossils from 56–54 mya.

Essentially, this site represents deposits laid down during the late Paleocene and early Eocene epochs, in a warm climate. The Paleocene rocks of the Thanet Formation are exposed on the foreshore and in the cliffs towards Reculver. The younger Paleocene and Eocene rocks overlay this and are exposed in the gently dipping strata and at Beltinge, the Beltinge Fish Bed of the Upnor Formation (Paleocene) is brought down to beach level.

EOCENE — YPRESIAN

PALEOCENE — THANETIAN

LONDON CLAY
HARWICH FORMATION
UPNOR FORMATION
THANET FORMATION

West of the car park, the Oldhaven Beds slope towards beach level, exposing the Oldhaven Fish Bed. Fish fossils (particularly shark teeth, ray teeth and crushing palettes and fish vertebrae) and turtle carapace are common and can be found in the shingle on the foreshore, preferably on a low spring tide.

Other fossils from the underlying Woolwich and Reading Beds can also be found but these are rarely exposed unless the low spring tide is favourable and the sea has retreated far enough.

The chances of finding teeth improve the further the tide goes out but fossils can generally be found at all times of the year, especially after stormy weather. Broadly speaking, the collecting area is in the section of beach between the groynes, either side of the concrete steps.

Careful work and a sharp eye in the pebbles and clay on the foreshore should reveal a good number of shark teeth, which are black in colour. Some are very small, so a pair of tweezers is a good idea!

The most productive area is immediately opposite the car park and for about 100 metres west. There are teeth from about 24 species of shark, ray and other fish to be found here, as well as the remains of crocodiles and turtles. Fossilised wood is common, although impossible to collect (as it turns to dust); pine cones can also be found.

The 'hunting ground' at Beltinge.

The crushing palette of an eagle ray.

Otodus obliquus, shark teeth from the Thanet Formation.

Stratiolamia macrota shark teeth from the Beltinge foreshore.

Site summary

Slippery rocks when wet

★★★★★ Accessibility
★★★★★ Child suitability
★★★★★ Find frequency

SSSI Site. No hammering or digging into the cliffs or bedrock.

Ideal for children.
Visit on a low tide.

Suggested equipment: Specimen bags, tweezers.

Directions: Head to Beltinge from the A299. Park in the car park at the end of Reculver Drive, where a concrete path leads down to the beach.

Postcode: CT6 6QE

In the east, towards Reculver, the Thanet Formation is sometimes revealed, provided the overlying beach sand is not present. The Thanet Formation can contain a good range of bivalve and gastropod shells.

ABBEY WOOD, LONDON SE2

SSSI Site – adherence to fossil collecting conditions essential

Abbey Wood is a site of geological interest set within a southeast London park at Lesnes Abbey Woods in the London Borough of Bexley. It is famous for the profusion of shark teeth, fish bones and scales and shells. Also remains of turtles, crocodiles, birds and, more rarely, small mammal teeth can be found.

The beds were laid down approximately 54.5 mya in a shallow tropical sea and contain a mixture of marine and estuarine, as well as freshwater fossils. The public can dig here, with prior permission from the Lesnes Abbey ranger. The site is late Palaeocene and early Eocene and comprises of the Lesnes Shell Bed of the Blackheath Member. The Blackheath Member rests on strata ranging from the lower part of the Upnor Formation to the lower part of the Thanet Formation.

EOCENE — YPRESIAN — HARWICH FORMATION

PALEOCENE — THANETIAN — THANET SAND FORMATION

OLDHAVEN MEMBER
BLACKHEATH MEMBER
UPNOR FORMATION
THANET SAND

The Abbey Wood site gives permission for visits via the Parks and Open Spaces department at Bexleyheath Civic Offices. There are conditions to collecting here and these should be observed by all.

You may remove no more than 2 kgs of material from the site. The park also asks that visitors do not dig to more than 2 feet in depth and ask that you refill any large or deep holes.

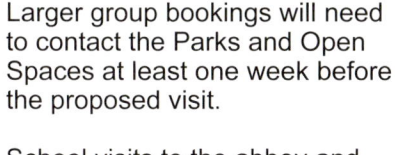

Larger group bookings will need to contact the Parks and Open Spaces at least one week before the proposed visit.

School visits to the abbey and the fossil beds are welcomed, but again, they should be booked in advance, so that the staff can aim to ensure toilet facilities are open.

Finding the fossils requires no skill. Simply dig a hole in a designated zone and the profusion of shark and ray teeth to be found is simply staggering.

Palette of an extinct wrasse, Phyllodus toliapicus.

Site summary

★★★★★ Accessibility

★★★★★ Child suitability

★★★★★ Find frequency

SSSI site. Observe the guidance rules for visits to this site.

Suggested equipment: Sieve, small trowel, specimen bags.

Apply for permission via Parks and Open Spaces, Civic Offices, 2 Watling Street, Bexleyheath, Kent DA6 7AT or email parksandopenspaces@bexley.gov.uk Tel: 020 8303 7777

Directions: By road, take the A206 Woolwich to Dartford then the B213 Abbey Wood to Belvedere. Parking on Abbey Road, Belvedere. By rail, the nearest railway station is Abbey Wood.

Postcode: DA17 5DL

A 30-minute search at Abbey Wood by Sam Caethoven produced these shark teeth (above) and shells (right).

BURNHAM-ON-CROUCH, ESSEX

SSSI Site – no hammering or digging into the cliffs

Burnham-on-Crouch lies on the north bank of the River Crouch and is one of Britain's leading places for yachting. The London Clay banks here still provide the collector with a rich assemblage of fish taxa, almost exclusively of elasmobranchs (cartilaginous fish, such as sharks, rays and skates). The tidal river cliffs and foreshore exposures reveal a section through the marine rich London Clay, Division D. The main cliff, known as 'The Cliff' or Butts Cliff locally, forms a 2–3 metre-thick outcrop on the north shore, which contains the fish fauna (mostly of sharks) which wash out onto the foreshore.

The fauna here is similar, in terms of composition and age, to that of the London Clay Division D at Sheppey, although at Burnham the scattered remains of teeth and bones are recovered in a better state than similar remains recovered from Sheppey exposures. It is the type locality for several fish species, for example the shark *Weltonia burnhamensis* and the ray *Burnhamia daviesi*.

EOCENE   YPRESIAN   LONDON CLAY FORMATION

The London Clay at Burnham-on-Crouch has also yielded bird bones and crustaceans. However, the most important fossils are the bones of birds. This includes type specimens of two small species, *Coturnipes cooperi* (a game bird) and *Parvicuculus minor* (a proto cuculid – a form of cuckoo).

The site is of considerable value in expanding the limited knowledge of small Eocene birds species and avian evolution.

Finding fossils on the foreshore at Burnham-on-Crouch is a 'hands and knees' search, due to the small size of the majority of finds. The site is extremely muddy (often thick in places) and there is a considerable amount of broken glass on the beach. It is not advisable for young children to participate in looking for fossils here.

Fossils can be found in the shingle and larger specimens are often trapped in larger stones and boulders and debris on the shore. Look under seaweed, where fossils can become lodged, due to tidal action.

Large cementstone nodules from the London Clay are found here and resemble those found at Sheppey sites. However, at Burnham-on-Crouch the concretions are largely unfossiliferous when broken open.

The main cliff contains the fish fauna, which wash out onto the foreshore.

Shark teeth found at Burnham-on-Crouch: *Hexanchus agassizi* (top left), *Notorhynchus serratissimus* (bottom left), *Jaekelotodus trigonalis* (top right), *Squatina prima* (bottom right).

Shark teeth and vertebrae from the foreshore at Burnham-on-Crouch.

Site summary

⚠️ Muddy foreshore

★★★ Accessibility

★★★ Child suitability

★★★★ Find frequency

SSSI Site. No hammering or digging into the cliffs.

Crumbling cliffs. Thick muddy foreshore and broken glass makes this an unsuitable site for children.

Suggested equipment: Small steel hand pick, specimen boxes, wrapping material.

Directions: Head towards the marina or ferry car park in Burnham-on-Crouch, southwest of the town. It is quite a walk to the site. Go westwards along the coastal path to Creeksea Slipway, onto the foreshore and cliffs.

Nearest postcode: CM0 8UB

BARTON-ON-SEA, HAMPSHIRE

SSSI Site – no hammering or digging into the cliffs

The Barton Beds at Barton-on-Sea are famed for their fossils, particularly the gastropods. More than 600 different types of shells are known from here and they represent a time when Britain bathed in a climate not dissimilar to Spain today. Other common fossils include shark and ray teeth, turtle bones and carapace fragments, fish remains, scaphopods and bivalves.

The section between Barton-on-Sea and Highcliffe is highly fossiliferous and yields are very high. The safe, sandy foreshore make this an ideal location for children. Climbing the crumbling cliffs is not recommended but at beach level the clay slippages provide ample scope for building a fine fossil collection, even during hotter summer weather when little erosion of the cliffs has taken place.

- EOCENE
- BARTONIAN
- BARTON FORMATION

- BECTON SAND
- CHAMA SAND
- UPPER BARTON BEDS G–L
- MIDDLE BARTON BEDS C–F
- LOWER BARTON BEDS A1–B

The staggering number of gastropods and bivalves found at Barton-on-Sea makes this an exceptional collecting site. Top row (left to right) gastropods: *Volutospina ambigua, Clavilithes longates, Turricula rostrata, Volutospina athleta*; bottom row (Left to right) bivalves: *Chlamys carinata, Cardita sulcata*.

The Barton Clay exposures date to 40 Ma and are representative of a marine environment. The clay was deposited at the bottom of a warm marginal sea shelf, close to land during the Eocene epoch.

The Barton Clay here is about 30 metres thick and overlain by the Chama Sand and Becton Sand. Changes in the sea level are reflected in the various layers found within the Barton Clay and the ten beds indicate shallow, near shore and deeper, off shore conditions. Plant material, particularly wood, is not uncommon, indicating the nearby land during deposition.

Collection of the fossils requires little effort. Walking east from the car park, along the beach, the clay cliffs can be searched from Highcliffe to Barton-on-Sea and many superb fossils can be extracted by simply using a small steel pick, with which they can be pried from the clay. The eroded clays leave the fossils protruding and they are easy to spot.

The shells are best left to thoroughly dry after which they can be carefully brushed, using a dry toothbrush, to remove any clay or sand. Use a dilution of PVA and water at 1:3 and they will be ready for display. Similar treatment can be applied to bones and carapace fragments. Shark and ray teeth require only a wash to remove any clay.

A haul of gastropods and bivalves from a UKAFH hunt at Barton-on-Sea in 2015.

A large gastropod *Clavilithes macrospira*.

The beach and cliffs at Barton-on-Sea.

Fish, ray and shark bones and teeth are a common find at Barton-on-Sea, including top row (left to right): *Cybium excelsum*, *Notorynchus kempi*, *Myliobatis* sp., *Striatolamia macrota*; bottom row (left to right): carcharhiniform shark vertebra, *Striatolamia macrota* tooth.

The sandy beach at Barton-on-Sea can provide shark teeth at low tide.

A *Volutospina* gastropod, a common find and along with other shells can be carefully prised out of the clay.

Site summary

Rock falls

★★★★★ Accessibility

★★★★★ Child suitability

★★★★★ Find frequency

SSSI Site. No hammering or digging into the cliffs.

Crumbling cliffs. Safe sandy foreshore makes this a suitable site for children. Safe access to foreshore from car park at the Highcliffe end of the section. Nearby beach refreshments and toilets (open all year).

Suggested equipment: Small steel hand pick, specimen boxes, wrapping material.

Directions: Park next to the Cliffhanger Cafe at the Highcliffe Cliff Top car park in Wharncliffe Road and walk down the slipway to the beach.

Postcode: BH23 5DF

A high spring tide at Barton-on-Sea.

The foreshore at Minster-on-Sea allows relatively safe hunting for London Clay fossils, especially shark teeth, crab and lobster carapaces and fruits and seeds. Fossil fish also turn up, such as *Rhinocephalus planiceps* and *Sciaenurus bowerbanki* below, both found in septarian nodules.

MINSTER, ISLE OF SHEPPEY, KENT

SSSI Site – no hammering or digging into the cliffs

The beach at Minster-on-Sea, on the north side of the Isle of Sheppey in Kent, is a London Clay site of Paleocene age (48–47 mya). Fossils here are abundant and different to those found at Warden Point, which is probably a more popular, albeit hazardous, location. Minster-on-Sea is more suitable for children (with supervision) and fossils can be picked from the foreshore, either in the beach shingle as pyritised specimens or in phosphatic nodules.

In the shingle, search on hands and knees for a variety of fruits and seeds (especially the large fruit of the palm *Nipa*), fish teeth and bones, gastropods, bivalves, snake and turtle remains, lobster and crab and shark teeth. Younger children will need close superviion, as the extensive foreshore can be very muddy and tides can reach the base of the cliffs.

EOCENE   YPRESIAN   LONDON CLAY FORMATION

Many of the fossils found at Minster are preserved in pyrite. As explained in an earlier chapter, the preservation of pyrite is a difficult problem. In damp conditions it rapidly decomposes to oxides of iron and sulphuric acid. This destroys any associated calcite and damages bone. The large fruits found appear to be particularly susceptible to 'pyrite disease'.

Most attempts to preserve pyrite, are relatively ineffective. Thorough washing, neutralisation and storage in silicon fluid, or within an airtight container with silica gel crystals, appears to be the best currently available method for the amateur collector.

There is a wealth of phosphatised fossils at Minster-on-Sea. Phosphatic nodules contain crustacean carapaces, burrow systems and large vertebrates. The surface of most of the nodules is soft and can be prepared away to expose superbly preserved fossil remains.

Shark teeth and fish bones, require no preparation, other than desalination.

Minster-on-Sea's foreshore.

A selection of shark teeth, collected from along the foreshore.

Worn fish skull in a phosphatic nodule.

Site summary

★★★★★ Accessibility
★★★★★ Child suitability
★★★★★ Find frequency

SSSI Site. No hammering or digging into the cliffs.

A relatively safe site, with plenty of specimens to be found. Ideal for children, although younger children will need close supervision on muddy foreshore. Best on a low tide. Car parking within a short distance of the fossil collecting area.

Suggested equipment: Small steel point, specimen boxes.

Directions: From the M2, take the A249 then the A250 towards Minster. Then follow the B2008 towards Minster town centre. Head to the northern part of the town towards the promenade, parking at the southeastern car park and walking towards the cliffs by way of the promenade.

Postcode: ME12 2LT

Various *Nipa* palm fruits, seeds, shark teeth and lobsters in nodules from Minster-on-Sea.

The foreshore at Warden Point. Fossils are found in the gravel accumulations on the beach, or amongst the phosphatic and septarian nodules. The photo of the crab, *Zanthopsis leachi* and the lobster, *Hoploparia*, were kindly provided by Fred Clouter.

WARDEN POINT, ISLE OF SHEPPEY, KENT

SSSI Site – no hammering or digging into the cliffs or bedrock

Warden Point is on the northeast side of the Isle of Sheppey. It is situated on a section of the island where the cliffs expose over 50 metres (vertically) of London Clay of early Eocene age. The site is famed for its fossils, which occur in pyrite accumulations on the beach or in cement stone nodules along the foreshore, having been washed out of the cliffs.

The site is rich in fossils but the pyritisation of many, including large fruits of the palm *Nina*, seeds, twigs and wood, gastropods and brachiopods makes them prone to pyrite 'disease' and they will decompose. Crab and lobster carapaces and bones preserved in phosphatic nodules remain stable once prepped, along with shark and ray teeth and fish bones.

EOCENE YPRESIAN LONDON CLAY FORMATION

The foreshore is undoubtedly the best place to search for fossils. The exposures, although most often covered in silt, can extend for 300 metres seawards. The quality and diversity of the fossils to be found here is recognised internationally.

The coastline here erodes at an alarming rate and the slumped cliffs are highly unstable. Do not venture near the cliffs. In any case, fossils are rarely found in the cliffs and are almost impossible to spot. Instead, examine the accumulations on the beach, found alongside iron pyrite casts. Here, various fruits and seeds, gastropods, molluscs, twigs and wood, coral and worm tubes occur, preserved in iron pyrites and which have been sifted by the sea. They occur as pockets on the beach. The beach accumulations also contain shark teeth and fish vertebrae.

The best ways of finding these smaller specimens is to get down on your hands and knees and search through the pyrite and wave sorted gravel very carefully. The teeth are found within the fine gravel above the level of the pyrite.

Fish vertebra, small fish skulls, small crabs and ray plates are usually found within the pyrite accumulations as well. The pyrite is dull, whereas the fossils are shiny black. However, as described elsewhere, the fossils preserved in pyrites are susceptible to 'pyrite disease' and will probably deteriorate over time.

The rapidly eroding cliffs are dangerous and should be avoided. The mudflows are also an unsafe area from which to collect.

Searching the gravel and pyrite accumulations for shark teeth and vertebrae. Work slowly from low down on the beach towards the cliffs. When teeth are encountered others will be approximately at the same level.

Crab carapace in a phosphatic nodule.

Crab in a phosphatic nodule.

Two large, fused fish vertebrae from Warden Point.

On the other hand, seaward and away from the gravel and pyrite accumulations, pale-cream coloured phosphatic nodules can be found, washed out onto the beach and protruding above the clay. The nodules should be examined for fossils, which can be recognised as either black coloured bone or carapace. The phosphatic nodules may also contain fish or turtle remains but be warned ... not all phosphatic nodules contain fossils and you may spend several hours searching to no avail!

Phosphatic nodules can contain exquisite fossils, with incredible detail preserved. Preparation is relatively easy and the matrix can be carefully removed by scraping with a blade or with the use of an air pen. Do not use acid!

Cement stones or concretions are rare and so are fossils within them. They are notoriously difficult to prepare and yet, the fossils found within them include crabs, fish, turtles, crocodiles and nautili.

The 2-mile section between Warden Bay and Hen's Brook (Eastchurch Gap), which includes Warden Point, is a highly productive area for fossils but there is always the risk of finding absolutely nothing! It is tide dependent and winter months, with storms and high rates of erosion, are more favourable to the collector.

The Eocene London Clay here is from the Ypresian stage of around 52–51 mya, when the area of Kent lay beneath a warm, shallow sea and where the average annual temperature was around 23°C.

Life was abundant and the nearest large land mass, which was probably 30 miles away, was populated by tropical vegetation, fringed by a swamp-like environment, in which birds, mammals and insects flourished.

The flora and fauna found at Warden Point and along the coastal section here presents fossils that are representative of that palaeoenvironment.

www.Sheppeyfossils.com is a website by Fred Clouter and provides a highly recommended resource for the Lower Eocene (Ypresian) fossils of the north coastal section of the Isle of Sheppey.

Large *Otodus* shark vertebra. *Otodus* is an extinct genus of mackerel shark.

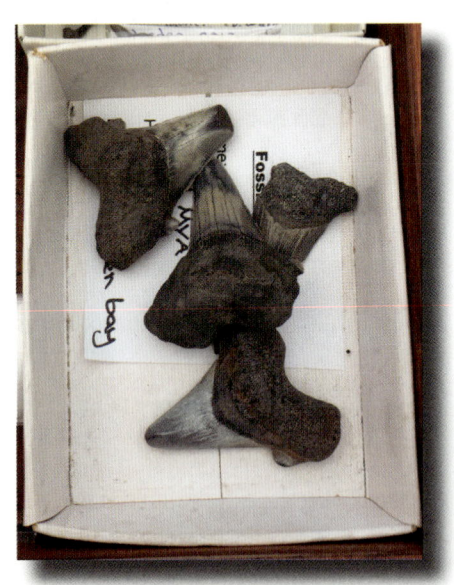

Large *Otodus* shark teeth.

Nipa palm fruit found by Sam Caethoven.

Fish in a phosphatic nodule.

Site summary

Deep mud

★★★★★ Accessibility
★ Child suitability
★★★ Find frequency

SSSI Site. No hammering or digging into the cliffs or bedrock.

A fairly hazardous site, with slippery and often deep mud. Fast currents and tidal. Not recommended for children. Car parking within a short distance of the fossil collecting area.

Suggested equipment: Small steel point, specimen boxes.

Directions: Access to Warden Point can be made on foot from Imperial Drive, Warden, Sheerness at the end of which a small gravelled car park provides free parking all year round.

Postcode: ME12 4SD

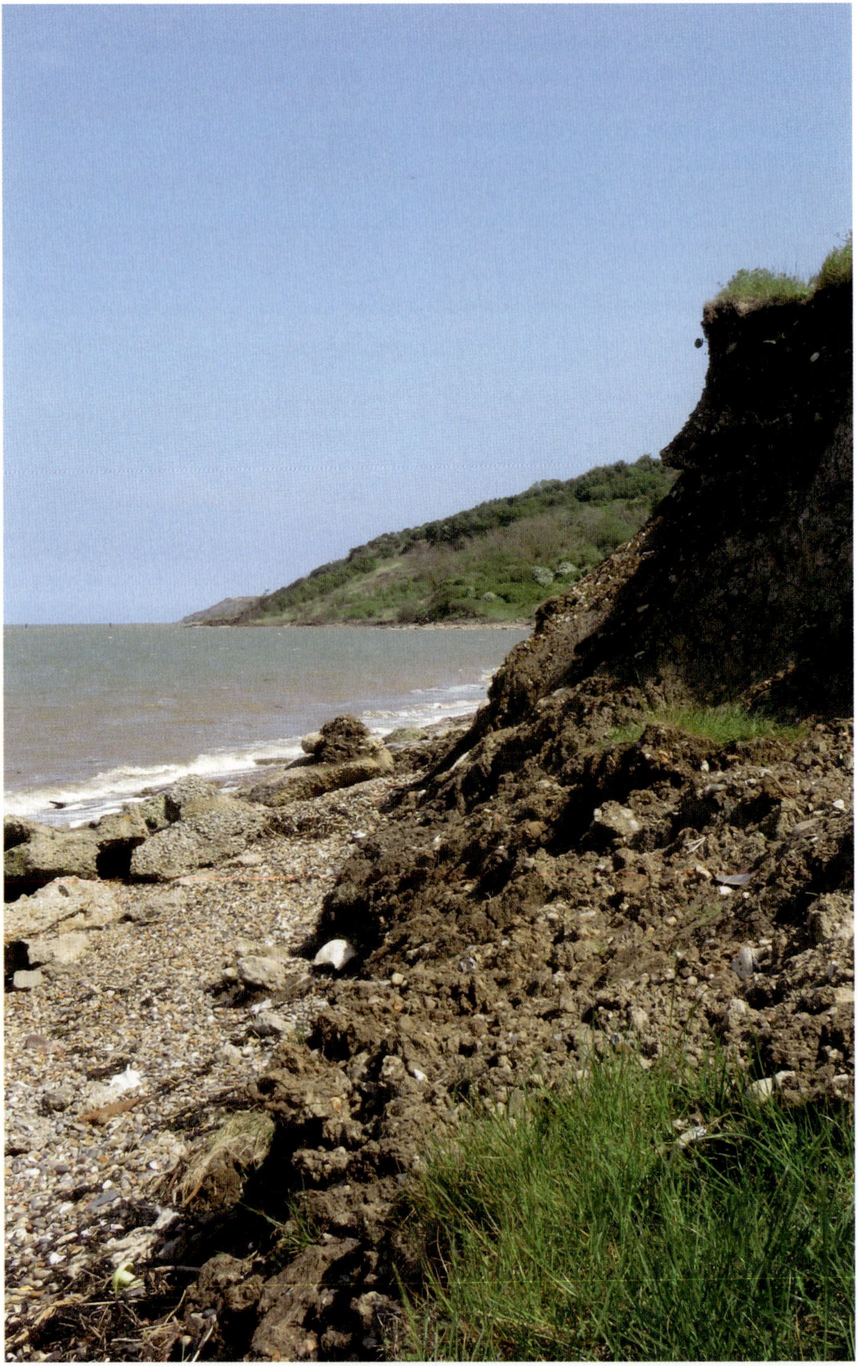

MAYLANDSEA, ESSEX

SSSI Site – no hammering or digging into the cliffs or bedrock

Maylandsea is situated on the Dengie peninsula of Lawling Creek in the Blackwater estuary of Essex. It forms part of the Blackwater SSSI. The London Clay Member present at this location is of Ypresian age (50 Ma) and outcrops on the banks and foreshore of the creek. The foreshore has yielded a vast number of fossils, particularly crustacean remains of lobsters and crab.

Despite the small size of the location, it can be productive with fish remains and shark teeth frequently occurring. Other fossils include seeds, bivalves, freshwater gastropods, as well as crinoid fragments.

EOCENE   YPRESIAN   LONDON CLAY FORMATION

The fauna found here represents the time of maximum transgression of the sea and includes the deposits of the deepest water. The faunal assemblage may well reflect this, with deeper-water species that are not seen at Sheppey or any of the other London Clay sites.

The lobsters are found in light yellow nodules and collection is best along the foreshore at low tide, on hands and knees. A focused, systematic serch of the beach shingle and gravels should reveal a number of fossils for collection. The phosphatic nodules are comparatively soft and fossils within them can be prepared relatively easily using hand tools, such as dental picks and a sharp blade.

Several species of lobster can be collected from Maylandsea, although *Hoploparia* is the most common find.

The beach and foreshore at Maylandsea.

Site summary

⚠️ Muddy site

★★★★★ Accessibility

★★★★ Child suitability

★★★★ Find frequency

SSSI Site. No hammering or digging into the cliffs or bedrock.

Suggested equipment: No equipment necessary. Specimens can be hand collected.

Directions: Head towards the west end of Maylandsea village and drive down North Drive, parking at the bottom. A footpath takes you along the banks of the River Blackwater. Walk eastward to the foreshore.

Postcode: CM3 6AG

Phosphatic nodules containing lobster remains.

The Bracklesham Group of clays contains a rich fauna of gastropods and bivalves including (top to bottom, left to right): *Ampullina grossa, Haustator* sp., *Venericor planicosta, Barbatia laekeniana, Volutospina nodosa* and teeth from the shark, *Striatolamia macrota*.

BRACKLESHAM BAY, WEST SUSSEX

SSSI Site – no hammering or digging into the bedrock

Bracklesham Bay is a fantastic site for the whole family. With its large sandy beach, easy parking and toilet facilities, it is a sure favourite for beginners and hardened hands. The location is highly productive and especially on low, retreating spring tides, when the fossil shell beds are revealed.

At most times of the year, fossils are deposited on the sand, being from the Lutetian stage (46 Ma) of the Bracklesham Group of sediments within the Hampshire Basin of West Sussex. Within the mid-Eocene clay beneath can be found a vast number of gastropods, bivalves, shark and ray teeth, foraminifera, coral, fish and turtle remains and other marine fossils.

EOCENE • LUTETIAN • BBACKLESHAM GROUP

WITTERING FORMATION
EARNLEY FORMATION
MARSH FARM FORMATION
SELSEY FORMATION

The shingle near to the car park is also worthy of searching, as many shark teeth end up in this section, having been washed up from the fossil beds further out.

A carapace fragment of the turtle, *Trionyx*.

An assortment of bivalves and gastropods washed out of the fossiliferous clays.

The clays at Bracklesham Bay are part of the Eocene Bracklesham Group. From the car park and approximately 1 km towards Selsey, the clays are of the Earnley Formation. These grey clays are exposed as 'mushroom'-shaped pedestals of clay, which are fully exposed on very low tides but even in less favourable conditions they will distribute their fossils on the sandy beach.

If you are able to get to Bracklesham Bay on a retreating, low spring tide, then these pedestals are impressively crammed with in situ fossils. Other beds of the Earnley Formation are also exposed, with staggering numbers of fossils to be seen and collected.

Beach conditions do vary and the site can occasional be disappointing, especially when covered in thick sand, but even under these conditions shells can be seen on the surface, particularly the large bivalve *Venericor planicosta*, the spiraled gastropod *Turritella* and the foraminifera *Nummelites*.

The fossil shell beds exposed on Bracklesham Bay beach at low tide.

Site summary

★★★★★ Accessibility
★★★★★ Child suitability
★★★★★ Find frequency

SSSI Site. No digging into the bedrock.

Suggested equipment: Collection boxes, wrapping materials.

Directions: Follow the A27 out of Chichester, heading west and follow the signs to Bracklesham Bay. Then take the A286, onto the B2201, then the B2145. The parking on the sea front car park is chargeable (pay & display). A toilet is within the car park and a refreshments kiosk (during the summer season). Once parked make your way down the beach area, directly in front of you.

Postcode: PO20 8JS

The foraminifera, *Nummelites*, which is a very common find at Bracklesham Bay.

The cliffs and foreshore at Walton-on-the-Naze are famous for their fossils from the Red Crag and London Clay. The orangey-red stained shells of the Red Crag are a familiar sight.

WALTON-ON-THE-NAZE, ESSEX

SSSI Site – no hammering or digging into the cliffs or bedrock

The rapidly eroding coastline at Walton-on-the-Naze is undoubtedly a fantastic location for collecting fossils, from both the London Clay and the directly overlying Red Crag Formation. The Red Crag was deposited 51 million years later than the London Clay and the sediment deposited in the interim (which included much of the London Clay) was eroded away. This, in part, contributes to the pebble bed at the base of the Red Crag. The London Clay is of early Eocene (Ypresian) age at approximately 54 Ma, whilst the Red Crag deposits are of Pliocene (Placenzian) age at around 2.5 Ma. Expect to find bivalves and gastropods in the Red Crag, with shark teeth, bird bones, twigs and wood, and fish remains in the London Clay.

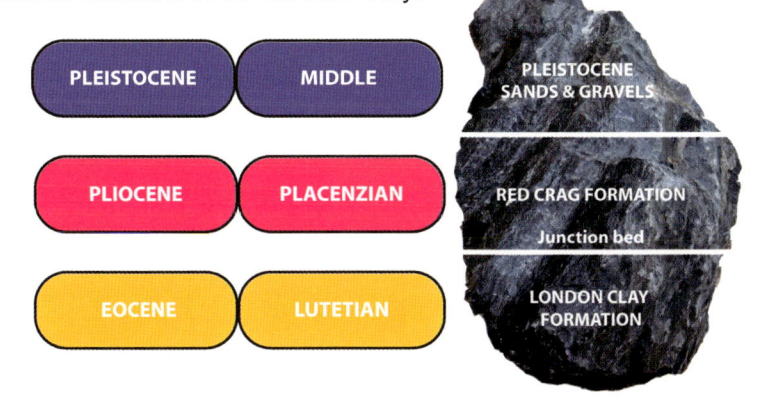

PLEISTOCENE	MIDDLE	PLEISTOCENE SANDS & GRAVELS
PLIOCENE	PLACENZIAN	RED CRAG FORMATION
		Junction bed
EOCENE	LUTETIAN	LONDON CLAY FORMATION

The bluish-grey coloured London Clay, of up to 170 metres (500 feet) in thickness, was laid down in the delta of a large river, in a subtropical climate and to the east of a large landmass, thought to have been about 250–300 kilometres to the southwest of the modern shoreline at Walton-on-the-Naze. The Eocene shoreline supported a subtropical rainforest, with plant species related to the mangroves, and the area probably closely resembled coastal regions now found in present day Indonesia and Malaysia.

Within the delta, rivers flowed into the sea bringing with them mud and silt, which eventually settled and compacted, forming the London Clay, now well exposed in the lower half of the cliffs at Walton-on-the-Naze.

Walton-on-the-Naze foreshore, beneath the London Clay and slumped Red Crag cliffs.

The London Clay at this location is highly fossiliferous and contains fossils of marine origin. The commonest of these are the teeth of the six-gilled sand shark (*Striatolamia macrota*). The larger teeth of the mackerel shark (*Otodus obliquus*) are also found and rarely the huge teeth of the giant shark *Carcharocles megalodon*.

Hands and knees sifting through the shingle on the foreshore is the best approach to finding shark teeth at Walton-on- the-Naze.

The clay also contains much plant material derived from rafts of material drifting out to sea from the heavily forested mainland, along with the bones of early mammals and birds that lived in the rainforest. Fossilised twigs and wood, preserved in iron pyrites, are common but are subject to pyrite decay and subsequent disintegration.

Tooth of the mackerel shark, *Otodus obliquus*.

The most common Red Crag fossils are bivalves and gastropods and nearly 300 species have been recorded from the Naze.

Shark tooth, bone fragment and whale vertebra.

At the very end of the Pliocene period (about 3 Ma), Essex was probably entirely covered by a sea between 15 and 25 metres (50 to 80 feet) deep, which deposited a sand bank close to the shoreline. An abundance of marine shells was laid down in these 'dunes' on the sea bed, fairly close to the shoreline. These shell banks now form the Red Crag, which can be seen resting on the London Clay.

The rust-red colour of the sand and its fossils is due to the former presence of iron pyrite, which was washed from the London Clay into the basal Red Crag and there oxidised. The product of this chemical reaction is a red deposit of iron oxide, which has stained the shells. Slumped masses of shelly sand are continually sliding down, to be removed rapidly by waves lapping against the foot of the cliff.

The junction between the London Clay and the Red Crag represents a time interval of about 47 million years and a period of erosive conditions, rather than sediment accumulation. The thin layer of phosphatic nodules found here contains fossils derived from the London Clay and well rounded lumps of sandstone, called 'boxstones', believed to be of Miocene age (about 10 Ma).

The Nodule Bed is often buried beneath the slumped cliffs but yields very rolled but highly polished bones of whale and large terrestrial mammals such as deer and elephant.

There is always the chance of finding a huge tooth from the shark *Carcharocles megalodon* on the beach at Walton-on-the-Naze.

A selection of finds from Walton-on-the-Naze, with shark teeth, fish vertebra, pyritised twigs and bird bone from the London Clay.

Carcharocles megalodon tooth. Not a common find but certainly the discovery of one is a 'lure' to many fossil collectors at Walton.

Site summary

★★★★★ Accessibility

★★★★★ Child suitability

★★★ Find frequency

SSSI site. No digging in the cliffs.

Find frequency variable. Best in winter months.

Easy access to the beach, via a series of steps and ramps from the Naze cliff top.
Pay & Display car park.
Toilets and cafe.

Suggested equipment: Steel point, collection boxes, wrapping material for the more fragile finds.

Directions: Heading north through the town, follow the left hand fork in the road into Hall Lane, then into Naze Park Road. Take the left hand road into Oldhallen Road, to the large car park. A cafe is inside the 'Naze Tower'.

Postcode: CO14 8LD (The Naze car park)

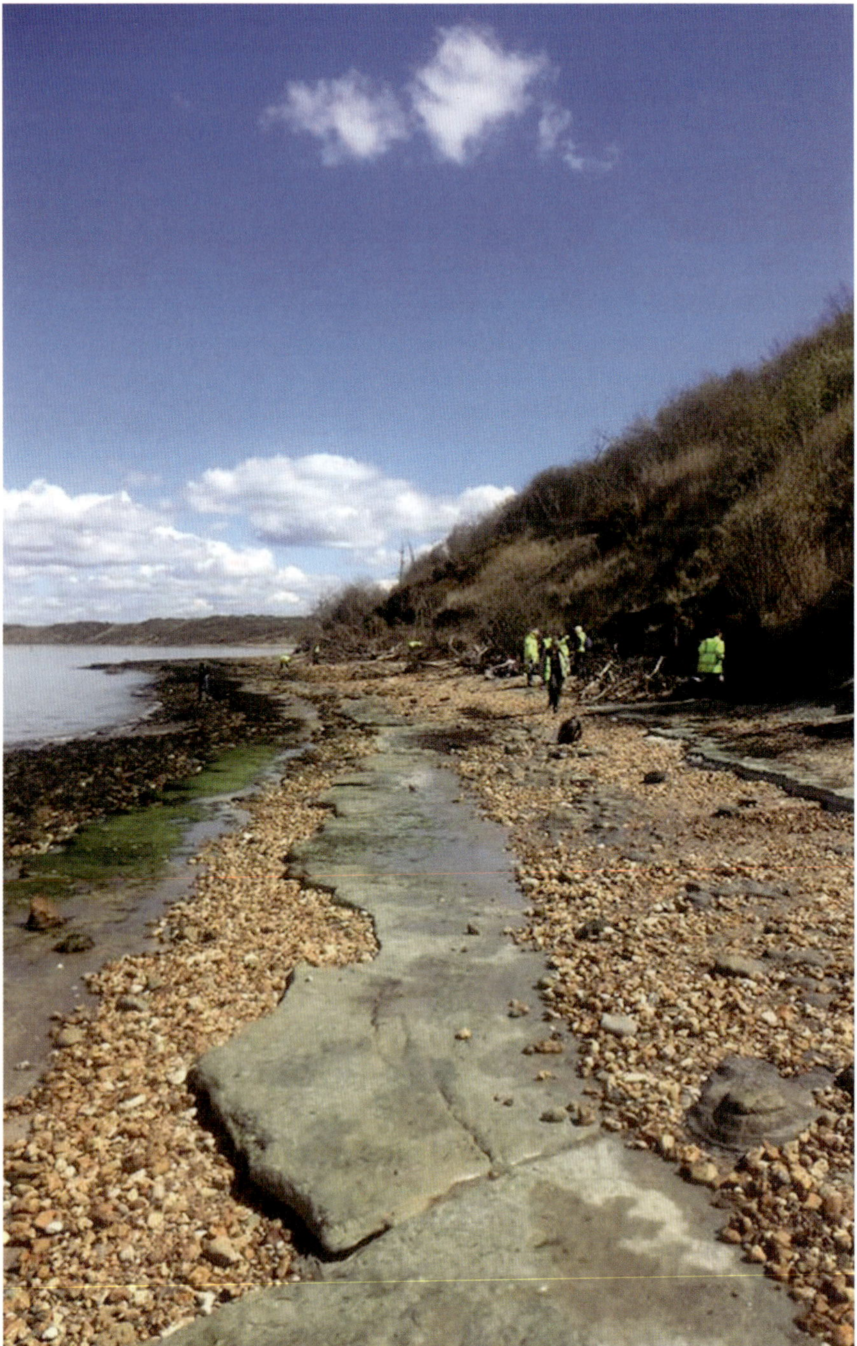

THORNESS BAY, ISLE OF WIGHT

SSSI Site – no hammering or digging into the cliffs or bedrock

Thorness Bay is situated on the north side of the Isle of Wight, near Gunard. The coast here is of Oligocene and Eocene origin. Only the northern and southern end of the bay are fossiliferous, the middle section being slipped Holocene deposits.

The Oligocene Bouldnor Formation is present but only the lower part, being the Hamstead Member. The Eocene Bembridge Marls Member, of upper Priabonian age (34–33.75 mya) also starts here below beach level and blocks of the famous Bembridge Insect Bed may be observed, especially during scouring conditions, although finds within this bed can be difficult because over-collecting has made the site less productive overall.

OLIGOCENE — RUPELIAN

EOCENE — PRIABONIAN

CRANMORE MEMBER
UPPER HAMSTEAD MEMBER
LOWER HAMSTEAD MEMBER
BEMBRIDGE MARLS MEMBER
BEMBRIDGE LIMESTONE

Foreshore collecting at Thorness Bay.

Section of crocodile jaw and teeth.

Turtle carapace found at Thorness.

This site requires patience and a hands and knees approach to finding fossils along the foreshore section at the north and south ends of the bay. The Eocene Bembridge Insect Beds can be identified by their bluish to green-grey laminated muds, with blue and buff coloured limestone and marl within it. Insect specimens are rare nowadays but in the past, more than 200 species of insects have been found here. However, within the bed are several horizons of molluscs and the fossil content represents a varied and rich biota, also including plants and vertebrates.

The Bembridge Marls, which is found along the foreshore, is a productive area for a number of fossils. Crocodile bones and teeth and turtle carapace fragments, some of a good size, are frequently found. The fossil content is quite varied, with both freshwater and marine shells, plant remains, seeds, arthropods and various vertebrate remains.

The overlying Oligocene beds of the Bouldnor Formation comprise the Lower Hamstead Member, followed by the Upper Hamstead Member. Along with various molluscs and gastropods, the beds are rich with Oligocene mammal remains from a large number of species.

Crocodile vertebra found at Thorness in 2015.

Tooth of the pig-like mammal, *Bothriodon*, from the Upper Hamstead Member at Thorness Bay.

Site summary

⚠️ Mud, take care

★★★★ Accessibility

★★★★ Child suitability

★★★ Find frequency

SSSI Site. No hammering or digging into the cliffs or bedrock.

Suggested equipment: Little or no equipment required. A gelogical hammer might be useful for breaking blocks of the Bembridge Insect Bed, if found.

Directions: Follow the coastal road (Marsh Road) from Cowes to Gurnard. From here, walk southwest along the beach, past Gunard Point to Thorness Bay. Alternatively access the beach via Thorness Bay Holiday Park, but the road to the park is private.

Postcode: PO31 8NJ (Thorness Bay Holiday Park)

The cliffs at Easton Bavents with the molar of the Southern mammoth, *Mammuthus meridionalis*.

EASTON BAVENTS, SUFFOLK

SSSI Site – no hammering or digging into the cliffs or bedrock

Easton Bavents is a Pleistocene site located on the coast of Suffolk and rates as one of the better locations for finding mammal remains. It is the only publicly accessible site for in situ mammal remains from the Norwich Crag and is recognised as being of international importance, although finds are generally limited to favourable tides.

The site has been subject to extensive erosion and as a result, the once productive cliffs are now being washed away at an alarming rate. The Norwich Crag has been subdivided into the Chillesford Sand and Chillesford Clay members, the latter of which crops out in the cliffs at Easton Bavents as a series of interbedded clays, sands and gravels of Baventian age.

PLEISTOCENE   BAVENTIAN   NORWICH CRAG   CHILLESFORD CLAY

Rough winter or spring tides are required to find washed out mammal bones and teeth.

Due to SSSI rules, foreshore collection is required and is the most productive way to find bones.

Cliff face showing the exposed shell bed but collection has to be limited to slippages.

Due to the high rate of erosion, remains of vertebrate fossils are regularly found on the foreshore, particularly after storms and spring tides. Less frequently, specimens have been found in situ in the cliff section.

Taxa collected from here include the proboscideans, *Mammuthus meridionalis* (or Southern mammoth) and *Anancus arvernensis* (an extinct genus of gomphothere), as well as horse (*Equus robustus*), deer (*Eucladoceros falconeri* and *Eucladoceros sedgwicki*), gazelle (*Gazella anglica*), beaver (*Trogontherium*), cetaceans (whale and dolphin), seals, walrus (*Alachatherium cretsii*) and wolves (*Canis cf. etruscus*). Other remains include those of water voles, birds, fish, shark and rays. The Norwich Crag shell beds also provide a rich fauna of molluscs at this site.

Fossils occur on the foreshore after high tides, during winter storms or during the period of spring tides. Waves wash them out of the cliffs, so collection is always best during such rough conditions.

Fossil shells, fish remains and bones can all be found in the cliff but collection is strictly limited to the slippages, due to the SSSI restrictions placed on in situ collecting and digging. The area beneath the smaller of the two cliffs at Easton Bavents is the most productive for mammal bones and teeth.

Molar of the elephant, *Anancus arvernensis*.

Part of the antler of deer, *Eucladoceros falconeri*.

Molar of the mammoth, *Mammuthus meridionalis*.

Site summary

Rock falls

Sinking sands after high tides

★★★★★ Accessibility

★★★★★ Child suitability

★★★ Find frequency

SSSI site. No hammering or digging into the cliffs.

Beware of tides. Collection best during winter and spring after storms and high tides.

Suggested equipment: Steel point, trowel, collection boxes and wrapping materials. Bones found fallen from the cliff or exposed in slippages can be fragile and should be either wrapped carefully in a towel and carried back for immediate preservation, or preserved onsite using a solution of PVA or similar.

Directions: From Southwold, drive to the car park in Pier Avenue. Walk to the foreshore to the second (smaller) cliff.

Postcode: IP18 6BN

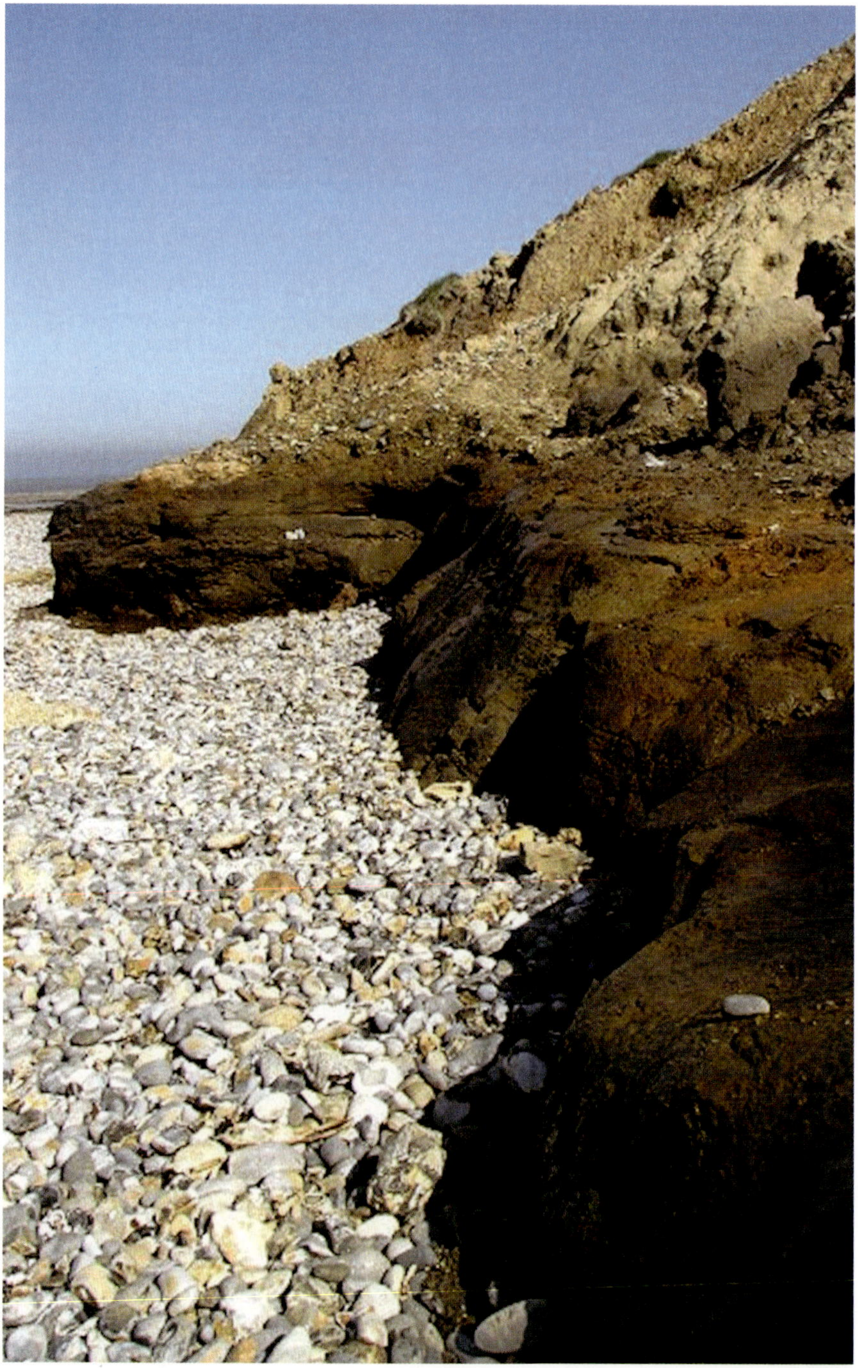

WEST RUNTON & EAST RUNTON, NORFOLK

SSSI Site – no hammering or digging into the cliffs

The Norfolk cliffs at West Runton to East Runton, just west of Cromer, provide a fantastic collecting opportunity from a number of beds and ages. Essentially, during the Ice Ages, glaciers to the north pushed through the land and these glacial deposits, along with Cretaceous Chalk, form the geology of the location.

The world-famous Upper Freshwater Bed is well exposed at the base of the cliff. These and the underlying sands, gravels and estuarine silts (Pastonian Stage) comprise the Cromer Forest-bed Formation, which can be exposed in the cliffs during scouring tides and rough weather.

- PLEISTOCENE
  - CROMERIAN
    - NORTH SEA DRIFT FORMATION
  - BEESTONIAN
  - CROMER FOREST-BED FORMATION
  - PASTONIAN
- CRETACEOUS

- GLACIAL TILL
- ESTUARINE & FRESHWATER SANDS
- WEST RUNTON FRESHWATER BED
- WROXHAM CRAG FORMATION
- CROMER STONE BED
- CHALK

Deer pelvis found by Mark Goble in 2014.

Typical fossil mammal bone found in the cliffs at West Runton.

The cliffs at East Runton, where the chalk 'rafts' are displayed, having been carried and shaped by the ice sheets.

The geology of the section between West and East Runton is complex. The beach itself is composed of flint, which sits on top of a Cretaceous Chalk wave-cut platform, with various erratics that were deposited by Ice Age glaciers.

Directly on top of the Chalk is the Cromer Stone Bed, followed by the Wroxham Crag. Old literature may refer to the Wroxham Crag as Weybourne Crag. The Wroxham Crag consists of sands, gravels, silts and clays, all of which represent a period of time from the pre-Pastonian cold stage, the Pastonian warm stage and into the Beestonian cold stage.

The Cromer Stone Bed and Wroxham Crag are often cemented together by calcrete and iron pan and are exposed during heavy tidal scouring conditions. The shelly sands, grey silts and conglomerates have produced a rich fauna, including voles, shrews, giant beaver, deer, elephant and it is significantly older than the West Runton Freshwater Bed.

Overlying the Wroxham Crag is the West Runton Freshwater Bed; a prominent dark 'peaty' bed at the base of the cliff and up to 2 metres thick. It is highly fossiliferous and contains the remains of plants and trees (seeds, cones, fungi and pollen), terrestrial and aquatic molluscs, fish (scales, teeth and bones), amphibians, large and small mammals (teeth and bones) and bird bones.

West Runton beach is most well known for the elephant or steppe mammoth *Mammuthus trogontherii*, which was discovered in 1990 and is one of the oldest fossil elephants to be found in the UK. Deposits of bone and teeth can still be collected from this bed but being an SSSI, it should not be dug into and it is better to collect from the beach, unless bone is obviously exposed at the surface.

The cliffs at West Runton, towards the east.

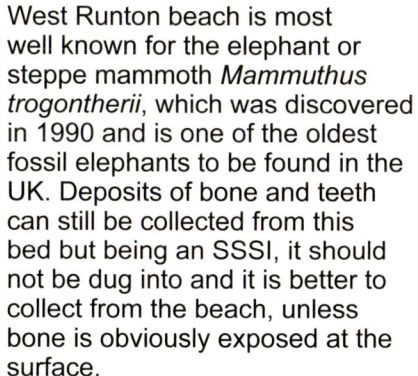

Whilst the main structure of the cliffs is made up of Pleistocene estuarine and freshwater sands, a Chalk foreshore allows for the collection of belemnites and sea urchins and other Cretaceous fossils.

The beach is strewn with massive flints and paramoudras, as well as Ice Age erratics. It is a great locality for a family fossil hunt and finds are commonplace, from the range of rock types featured here.

The peaty Cromer Forest-bed in which a partial skull and two sets of teeth of a prehistoric rhino, *Stephanorhinus hundsheimensis*, were found in January 2015 (below).

A small mammal jaw from the Freshwater Bed at West Runton.

Echinoid found in the Chalk at West Runton.

The West Runton 'elephant', now known to be a steppe mammoth, *Mammuthus trogontherii*.

Site summary

⚠️ Rock falls

⚠️ Slippery rocks

★★★★★ Accessibility

★★★★★ Child suitability

★★★★★ Find frequency

SSSI site. No hammering or digging into the cliffs.

Foreshore can be slippery. Beware of tides. Collection best on low, receding tide between East and West Runton. No beach lifeguards at West Runton.

Suggested equipment: Steel point, trowel, collection boxes and wrapping materials.

Directions: West Runton is located to the west of Cromer. There are car parks at both East Runton and West Runton gaps; cafe and toilets at West Runton beach, amenities in East Runton. The proprietor makes a small charge except in winter. Pay at the cafe if there is no attendant.

Nearest postcode: NR27 9ND (West Runton car park), NR27 9PA (East Runton, Beach Road car park)

RAMSHOLT, SUFFOLK

SSSI Site – no hammering or digging into the cliffs

Ramsholt is a river estuary location along the banks and foreshore of the River Deben and provides excellent collecting opportunities for finding fossils from the Red Crag, Coralline Crag and London Clay.

Derived fossils from the Red Crag Basement Bed include crabs, lobsters, fish remains and shark teeth, whereas the Coralline Crag provides a rich fauna of echinoids, corals, shells and bryozoans. The London Clay here is highly fossiliferous, with shark, fish, turtle, crab, lobster and snake remains to be found. The whole section is tidal and with the sea reaching the base of the cliffs, a regular supply of fossils are washed out, to be picked up along the foreshore.

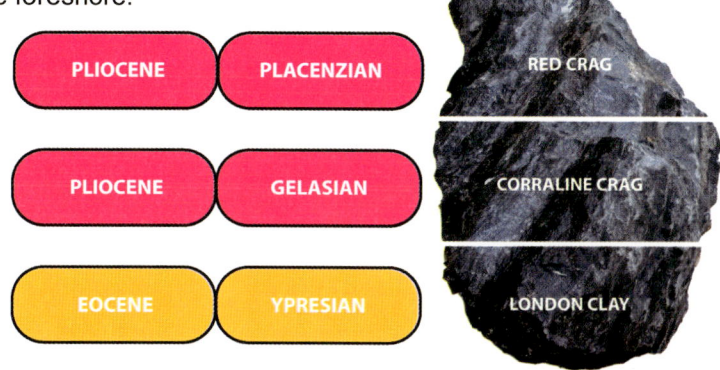

PLIOCENE	PLACENZIAN	RED CRAG
PLIOCENE	GELASIAN	CORALLINE CRAG
EOCENE	YPRESIAN	LONDON CLAY

A UKAFH hunt, about to commence at Ramsholt.

The western section, where fossils from the Coralline Crag are particularly prevalent.

The productive fossil collection area, under the trees on the foreshore.

This SSSI at Ramsholt reveals excellent exposures of a fascinating aspect of the Neogene Crags of East Anglia. Here the Pliocene rocks of the Coralline Crag, of Gelasian age (about 2.58–1.8 mya) are present, while the later Red Crag, of Placenzian age (about 3.6–2.58 mya) can be seen lapping over the Coralline Crag. The Coralline and Red Crags are the lowest two of the four formations within the Crag Group (the Norwich Crag and Wroxham Crag being the other two).

The isolated outlier of Coralline Crag found at Ramsholt is the most southern exposure in England and is highly fossiliferous. The western end of the section is best for shells from the Coralline Crag, whilst washed out shark teeth and ray plates accompany the mass of shells found here. The exposure of low cliffs nearest to the overhanging trees seems particularly rich in fossils from all three beds and yields a good number of shark teeth and crabs from the underlying London Clay of Ypresian age (56–47.8 mya).

The sheer number of fossils to be found here is quite overwhelming and the collector will certainly not be disappointed with this location. Please take note of the tides here, as the narrow foreshore can quickly become flooded!

An assortment of bryozoans from the Coralline Crag at Ramsholt.

Various shells from the Coralline Crag, including a large *Pectan*.

Shark teeth from the London Clay.

Site summary

★★ Accessibility

★★★★ Child suitability

★★★★★ Find frequency

SSSI Site. No hammering or digging into the cliffs.

Suggested equipment: Steel point, specimen boxes, wrapping materials (cotton wool, tissues, bubble wrap).

Directions: The site is not easy to find and requires a good 40 minute walk from the nearest car parking area. The walk along a footpath can be muddy in winter months. Park near the New Ramsholt Arms, Dock Road, Ramsholt, Suffolk. Walk along the riverside footpath north of the pub, for around 40 minutes, passing through the woods until the River Deben bends. Drop down to beach level and locate the area of low cliffs.

Postcode: IP12 3AB

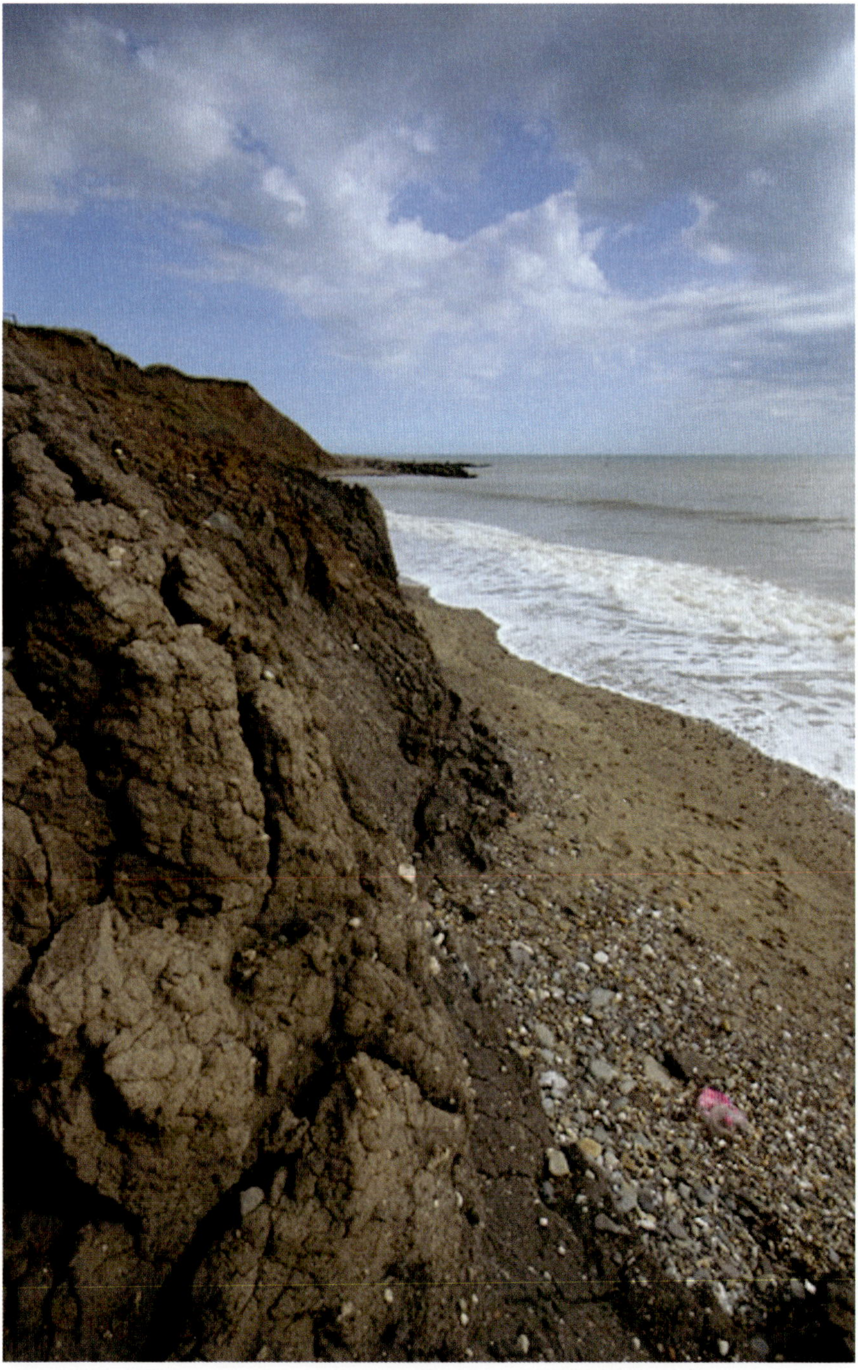

MAPPLETON, EAST RIDING OF YORKSHIRE

Mappleton is located on the Holderness coast of Yorkshire's East Riding. Here, the cliff line erodes at an alarming rate and the boulder clay and glacial till, deposited during the last period of Britain's Ice Ages, contains the rocks from other areas, as the glaciers pushed their way though Carboniferous, Jurassic and Cretaceous lands.

As a result, finds are plentiful but fossils found at Mappleton can range dramatically in size; from small belemnites to large pieces of oak that would have fallen when the ice sheets came down. It is one of those sites where just about anything might be found; ammonites, corals, belemnites and more. It's an ideal family fossil hunting location and with a high frequency of finds, it is an ideal hunting ground for the beginner.

PLEISTOCENE **DEVENSIAN** **GLACIAL TILL & BOULDER CLAYS**

CHALK

CRETACEOUS

Glacial till (often referred to as Boulder Clay) is a soft rock, identified by large angular rock fragments on the surface and within the soil. These rock fragments are erratics and it is within these that the fossils are found. None of the fossils found at Mappleton originate from the area. The till is unsorted material transported by ice sheets and is unstratified. At Mappleton, the till deposited at the coast totally obscures the Cretaceous Chalk bedrock.

Because of the fast erosion, the find frequency is high. As is the case at most coastal locations, the sea will have done much of the work already, in separating the clay from the fossils and depositing them on the beach. Search the foreshore, in the clay and among the rocks. Fossils can also be found in the slumped cliffs. By splitting the rocks found within the clays, with a geological hammer, many fossils can be found and are often extremely well preserved.

Large pieces of oak wood bear testimony to the glacial erosion that took place, forging a path to the Holderness coast. Fossils are often found within this wood, which has been infilled with the boulder clay.

The glacial till cliffs at Mappleton.

The Holderness coast.

Site summary

★★★★★ Accessibility

★★★★★ Child suitability

★★★★ Find frequency

Ex-M.O.D. site near car park. Do not touch objects en route to the beach.

Tidal location, with very strong currents. Ensure to collect on a falling tide.

Suggested equipment: Geological hammer, collection boxes, wrapping materials.

Directions: Access to the beach at Mappleton is easy. Free parking at the cliff top car park and access via a slipway.

Postcode: HU18 1XG

The Jurassic fossils shown above (ammonites, belemnite and *Gryphaea* oyster) are typical of the finds that you are likely to make at Mappleton.

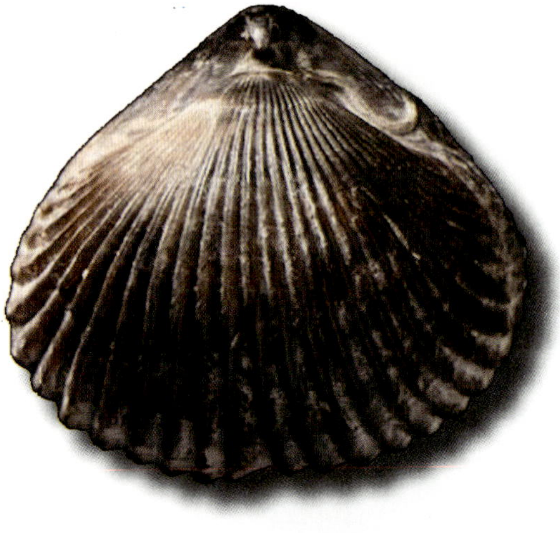

SECTION 5
GUIDE TO COLLECTING FOSSILS

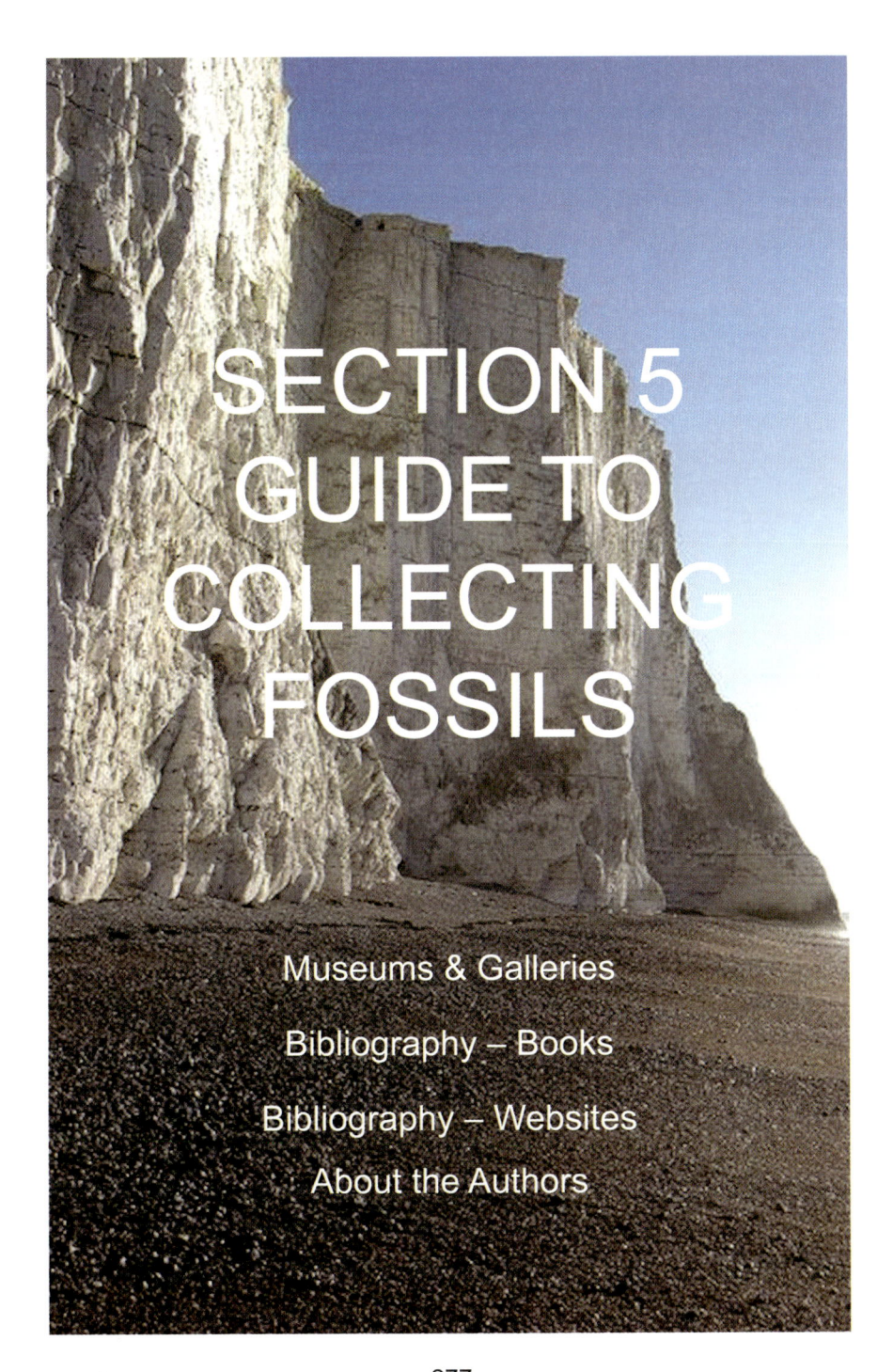

Museums & Galleries

Bibliography – Books

Bibliography – Websites

About the Authors

MUSEUMS & GALLERIES

Here are just a few recommended museums and galleries that have good collections of fossils or fossil related exhibits.

Natural History Museum, London is a museum that exhibits a vast range of specimens from various realms of natural history.
Cromwell Rd, London SW7 5BD

The Oxford University Museum of Natural History, sometimes known simply as the Oxford University Museum.
Parks Rd, Oxford OX1 3PW

Manchester Museum has displays of archaeology, anthropology and natural history (including a dedicated palaeontology gallery) and forms part of the University of Manchester.
Oxford Rd, Manchester M13 9PL

National Museum Cardiff is a museum and art gallery in Wales, with exhibitions ranging from contemporary art to dinosaurs, including the recently discovered Welsh dinosaur *Dracoraptor hanigani*.
Cathays Park, Cardiff CF10 3NP

The Lyme Regis Museum is a local museum in the town of Lyme Regis on the Jurassic Coast in Dorset. The museum building houses local fossils and those of Elizabeth Philpot.
Bridge St, Lyme Regis DT7 3QA

The Sedgwick Museum of Earth Sciences, is the geology museum of the University of Cambridge. A quaint, traditional museum with wooden cases and drawers full of fossils.
Downing St, Cambridge CB2 3EQ

Dinosaur Isle is a purpose-built dinosaur museum located on the Isle of Wight.
Culver Parade, Sandown PO36 8QA

New Walk Museum, Leicester. The Dinosaur Gallery: Exploring Lost Worlds. Suitable for all ages, the gallery includes interactives, hands-on activities and reconstructions of marine reptiles.
53 New Walk, Leicester LE1 7EA

Bristol Museum & Art Gallery. Archaeology, geology and art, including fossils.
Queens Rd, Bristol BS8 1RL

Whitby Museum is a private museum in Whitby, North Yorkshire. It contains one of the finest collections of Jurassic fossils in Britain, with specimens collected both in the early nineteenth century and from the present day.
Pannett Park, Whitby YO21 1RE

The Etches Collection is a museum of Jurassic marine life and the result of one man's passion. Steve Etches has devoted 30 years to the discovery and research of more than 2000 specimens he found in the Kimmeridgian Clay Formation, which are now housed in an impressive museum.
Kimmeridge, Wareham BH20 5PE

BIBLIOGRAPHY – BOOKS

The authors recommend the following for further reading and which were used as reference resources in the writing of this book.

Allison, R.J. The Coastal Landforms Of West Dorset. [London]: Geologists' Association, 1992.
Arkell, W.J. The Geology Of The Country Around Weymouth, Swanage, Corfe & Lulworth. London: H.M. Stationery Office, 1947.
Batten, D. English Wealden Fossils. London: Palaeontological Association, 2011.
Bone, D.A. Fossil Hunting at Bracklesham and Selsey. Chichester: Limanda, 2009
British Palaeozoic Fossils. London: Natural History Museum, 2012.
British Mesozoic Fossils. London: Natural History Museum, 2013.
British Cenozoic Fossils. London: Natural History Museum 2016.
Chatwin, C.P. East Anglia And Adjoining Areas. London: H.M. Stationery Office, 1961.
Cleal, C.J. & Thomas, B.A. Plant Fossils of the British Coal Measures London: Palaeontological Association, 1996.
Cope, J.C.W. Geology Of The Dorset Coast. London: Geologists' Association, 2012.
Davies, G.M. Dorset: A Geological Guide. London: Thomas Murby & Co., 1935.
Dewey, H. South West England. London: H.M. Stationery Office, 1948.
Earp, J.R. The Welsh Borderland. London: H.M. Stationery Office, 1971.

Gallois, R.W. The Wealden District. London: H.M. Stationery Office, 1965.
Green, G.W. & Welch, F.B.A. Bristol And Gloucester Region. London: H.M. Stationery Office, 1992.
Hains, B.A. Central England. London: H.M. Stationery Office, 1969.
Insole, A., Daley, B. & Gale, A. The Isle of Wight. Geologists' Association Guide No. 60. London: Geologists' Association, 1998.
Kent, P.E. Eastern England. London: H.M. Stationery Office, 1981.
Lomax, D.R. & Tamura, N. Dinosaurs Of The British Isles. Manchester: Siri Scientific Press, 2014.
Lord, A.R. & Davis, P. Fossils From The Lower Lias Of The Dorset Coast. London: Palaeontological Association, 2010.
Melville, R.V, Freshney, E.C. & Chatwin, C.P. The Hampshire Basin. London: H.M. Stationery Office, 1982.
Rawson, P.F. & Wright, J.K. The Yorkshire Coast. Geologists' Association Guide No. 34. London: Geologists' Association, 1996.
Rayner, D., Mitchell, T., Rayner, M. & Clouter, F.H. London Clay Fossils Of Kent And Essex. Rochester: Medway Fossil and Mineral Society, 2009.
Robinson, E. The Geology of Watchet and Its Neighbourhood, Somerset. Geologists' Association Guide No. 66. London: Geologists' Association, 2006 (revised 2011).
Ruffell, A. Early Cretaceous Environments of The Weald. Geologists' Association Guide No. 55. London: Geologists' Association, 1996.
Sherlock, R.L. London And Thames Valley. London: H.M. Stationery Office, 1962.
Smith, A.B. & Batten, D.J. Fossils Of The Chalk. London: Palaeontological Association, 2002.
Stone, P., Millward, D. & Young, B. Northern England. London: H.M. Stationary Office, 2010.
Swinton, W.E. Fossil Amphibians And Reptiles. London: Printed by order of the Trustees of the British Museum (Natural History), 1962.
Thackray, J. British Fossils. H.M. Stationary Office for the British Geological Survey, 1984.
Young, J.R., Gale, A.S., Knight, R.I. & Smith, A.B. Fossils Of The Gault Clay. London: Palaeontological Association, 2010.

BIBLIOGRAPHY – USEFUL WEBSITES

The authors recommend the following websites for further reading and which were used as reference resources in the writing of this book:

www.ukfossils.co.uk by Alister Cruickshanks – an online guild to the UK's fossil sites.
www.discoveringfossils.co.uk – by Roy Shepherd. A collection of online resources and guided fossil hunts that introduce the palaeontology of Great Britain.
www.jurassiccoast.org – Jurassic Coast.
www.Sheppeyfossils.com – by Fred Clouter. A guide to Lower Eocene (Ypresian) fossils of the north coastal section of the Isle of Sheppey.
www.gaultammonite.co.uk – a guild to fossils of the Lower Cretaceous Albian (Gault Clay and Folkestone Beds) in the county of Kent.
www.chalk.discoveringfossils.co.uk – by Robert Randell. A guide to British Chalk fossils.
www.southampton.ac.uk/~imw/index.htm – by Ian West. Geology of the Wessex coast of southern England.
www.dmap.co.uk/fossils/ – by Alan Morton. A collection of Eocene and Oligocene fossils.
http://www.trg.org/downloads/fossils%20of%20abbey%20wood.pdf – Abbey Wood fossils.
www.ukge.com – The market leader for geological and expedition equipment.

www.depositsmag.com – Europe's number one fossil and geological magazine.
www.palass.org – The website of The Palaeontological Association.
http://jncc.defra.gov.uk/pdf/gcrdb/GCRsiteaccount2056.pdf – The Woodhill Bay Fish Bed at Kilkenny Bay, Portishead.
http://jncc.defra.gov.uk/pdf/gcrdb/GCRsiteaccount1617.pdf – Tites Point (Purton Passage).
http://jncc.defra.gov.uk/pdf/gcrdb/GCRsiteaccount1976.pdf – Seaham.
http://jncc.defra.gov.uk/pdf/gcrdb/GCRsiteaccount220.pdf – Hunstanton cliffs.
http://jncc.defra.gov.uk/pdf/gcrdb/GCRsiteaccount2914.pdf – Burnham-on-Crouch.
www.siriscientificpress.co.uk – specialist UK-based publisher of palaeontology books.
www.ukafh.com – The UK's number one family friendly fossil club.

UK AMATEUR FOSSIL HUNTERS

THE UK'S LARGEST FAMILY FRIENDLY FOSSIL HUNTING GROUP WITH NATIONAL HUNTS & EVENTS EVERY MONTH

- AIMED AT THE AMATEUR ENTHUSIAST OR COLLECTOR
- PROFESSIONALLY ORGANISED
- FUN & EDUCATIONAL
- GREAT LOCATIONS CAREFULLY SELECTED

Join today - sign up at www.ukafh.com

Membership starts at as little as £12 a year allowing access to fossil hunting events and guaranteed success in finding fossils to take away at each location visited

As seen on ITV's 'Dinosaur Britain'

Registered students 10% Discount
UKGE limited
Established in 1998, we now have over 11,000 products online. Worldwide shipping.
• Supplier of geological and microscopy products
• Supplying educational establishments throughout the UK, Europe and World.
• Geological maps, books, chartwell
• Rocks and Minerals
• Microscopy and Science equipment.
ROCKS
FOSSILS
GEOLOGY
UK FOSSILS NETWORK
Deposits Magazine
Deposits Magazine is suited to a wide audience, including articles aimed at professionals, enthusiasts and beginners, covering international topics on rocks, fossils and minerals.
www.depositsmag.com
UK: 0800 0336 002
International: +441502 725205
www.ukge.com
Trade accounts welcome for educational establishments, simply Fax your purchase orders to +44(0)1502 725447
Suppliers to education, organisations and business
We also serve the general public and offer wholesale products

ABOUT THE AUTHORS

Steve Snowball

Steve began collecting fossils from the age of eight. He spent 34 years in teaching, eventually retiring from his post as a primary school headteacher to live on the Jurassic Coast of West Dorset. Here he divides his time between his role as an Ambassador for the Jurassic Coast Trust, as the Deputy Head of UKAFH, walking his dogs and following his lifelong hobby of collecting fossils. Steve is married with two children.

Craig Chapman

Craig is the founder of UKAFH. His keen interest in fossils led to the start of a local fossil collecting club, which has since grown to become the UK's largest and most popular fossil hunting organisation. Craig lives in the East Midlands, near to some fossil-rich localities, which helps him add to his personal, extensive fossil collection, especially of Pleistocene mammal remains. Craig is married with two children.

RELATED TITLES FROM SIRI SCIENTIFIC PRESS

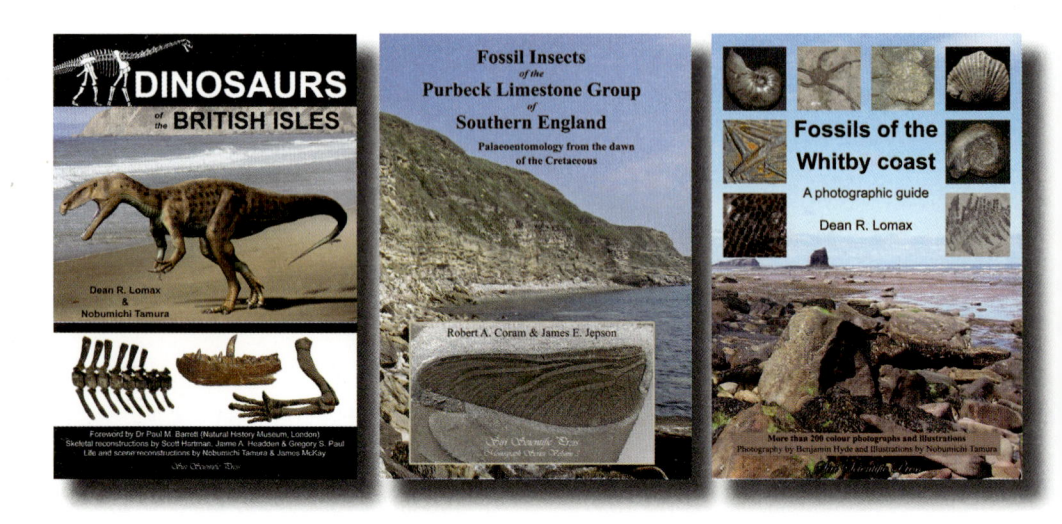

Visit our website (www.siriscientificpress.co.uk) for a diverse range of palaeontology titles and follow our Facebook page for regular updates on new and forthcoming titles.